ENERGY SCIENCE, ENGINEERING AND TECHNOLOGY

PRODUCING OIL AND NATURAL GAS FROM SHALE

ECONOMIC AND BUDGETARY EFFECTS

ENERGY SCIENCE, ENGINEERING AND TECHNOLOGY

Additional books in this series can be found on Nova's website under the Series tab.

Additional e-books in this series can be found on Nova's website under the e-book tab.

ENERGY SCIENCE, ENGINEERING AND TECHNOLOGY

PRODUCING OIL AND NATURAL GAS FROM SHALE

ECONOMIC AND BUDGETARY EFFECTS

JOELLE BOLTON
EDITOR

nova publishers
New York

Library of Congress Cataloging-in-Publication Data

ISBN: 978-1-63482-126-1

Published by Nova Science Publishers, Inc. † New York

CONTENTS

PREFACE

This book discusses the economic and budgetary effects of producing oil and natural gas from shale. It also examines the production, infrastructure, and market issues in U.S. shale gas development; and potential budgetary effects of immediately opening most federal lands to oil and gas leasing.

Chapter 1 – Recent advances in combining two drilling techniques, hydraulic fracturing and horizontal drilling, have allowed access to large deposits of shale resources—that is, crude oil and natural gas trapped in shale and certain other dense rock formations. As a result, the cost of that "tight oil" and "shale gas" has become competitive with the cost of oil and gas extracted from other sources. Virtually nonexistent a decade ago, the development of shale resources has boomed in the United States, producing about 3.5 million barrels of tight oil per day and about 9.5 trillion cubic feet (Tcf) of shale gas per year. Those amounts equal about 30 percent of U.S. production of liquid fuels (which include crude oil, biofuels, and natural gas liquids) and 40 percent of U.S. production of natural gas. Shale development has also affected the federal budget, chiefly by increasing tax revenues.

The production of tight oil and shale gas will continue to grow over the next 10 years—by about 30 percent and about 60 percent, respectively, according to a recent projection by the Energy Information Administration (EIA). Another EIA estimate shows that the amount of tight oil and shale gas in the United States that could be extracted with today's technology would satisfy domestic oil consumption at current rates for approximately 8 years and domestic gas consumption for 25.

Chapter 2 – The United States is in the process of discovering the true extent of its unconventional natural gas resources primarily derived from shale, with estimates rising every year. Natural gas has become an increasingly available, low-priced commodity, and U.S. natural gas is in demand at home and abroad. The price of natural gas relative to coal has led to a trend of switching from coal to gas for electric generation.

Estimates of undiscovered technically recoverable natural gas have multiplied in recent years, rewriting the map on natural gas production. Pennsylvania is newly one of the top four gas-producing states (since 2011), with Texas holding the top spot. This rise has necessitated pipeline network expansions; however, progress has been slow, leading to constraints and price differentials. Appalachian natural gas has historically been priced higher than that from the Gulf Coast's Henry Hub due to lower availability, but increased production in Pennsylvania's Marcellus Shale resource has reversed this differential.

The earning potential for liquefied natural gas (LNG) exportation overseas, particularly in high-demand areas such as Asia, has led to an increase in applications to construct U.S. export terminals. The prospect of increased LNG exports raises three main issues. First, exporting may raise domestic prices just as domestic natural gas usage is beginning to expand. Second, an influx of LNG into the global market may create new geopolitical concerns. Natural gas production in the United States has prompted other countries to attempt to produce from shale deposits. Third, onshore export terminals may be built in ecologically sensitive areas with the alternative being offshore terminals. Additionally, an increased reliance on LNG may shift focus away from exploring renewable energy sources.

The increase in U.S. natural gas production also affects the domestic market. The percent of new housing constructed to be heated by natural gas has become almost equal to the percent heated by electricity. This increase has been due largely to economic and environmental considerations. In the transportation sector, compressed natural gas (CNG) and LNG are beginning to displace traditional liquid fuels; some companies are producing vehicles that run on various forms of natural gas. However, the main obstacles are the cost of the new technology required and that of reformatting fueling stations.

Chapter 3 – The federal government offers private businesses the opportunity to bid on leases for the development of on- and offshore oil and natural gas resources on federal lands—although not all federally controlled lands are open to leasing now. The Congressional Budget Office (CBO) estimates that, under current laws and policies, the government's gross proceeds from all federal oil and gas leases on public lands will total about $150 billion over the next decade.

CBO has analyzed a proposal to immediately open most federal lands to oil and gas leasing, which would affect the amounts the federal government collects in various fees and royalties both in the near term and over a longer period.

In: Producing Oil and Natural Gas …
Editor: Joelle Bolton
ISBN: 978-1-63482-126-1

Chapter 1

THE ECONOMIC AND BUDGETARY EFFECTS OF PRODUCING OIL AND NATURAL GAS FROM SHALE[*]

Ron Gecan, Natalie Tawil and Mark Lasky

NOTES

All years referred to in this report are calendar years, not fiscal years. Numbers in the text and tables may not add up to totals because of rounding.

The photograph on the cover, "Rig," is attributed to Nicholas A. Tonelli. It comes from Flickr, where it was posted subject to the Creative Commons Attribution 2.0 license.

SUMMARY

Recent advances in combining two drilling techniques, hydraulic fracturing and horizontal drilling, have allowed access to large deposits of shale resources—that is, crude oil and natural gas trapped in shale and certain other dense rock formations.[1] As a result, the cost of that "tight oil" and "shale gas" has become competitive with the cost of oil and gas extracted from other

---

[*] This is an edited, reformatted and augmented version of a report issued by the Congressional Budget Office, December 2014.

sources. Virtually nonexistent a decade ago, the development of shale resources has boomed in the United States, producing about 3.5 million barrels of tight oil per day and about 9.5 trillion cubic feet (Tcf) of shale gas per year. Those amounts equal about 30 percent of U.S. production of liquid fuels (which include crude oil, biofuels, and natural gas liquids) and 40 percent of U.S. production of natural gas. Shale development has also affected the federal budget, chiefly by increasing tax revenues.

The production of tight oil and shale gas will continue to grow over the next 10 years—by about 30 percent and about 60 percent, respectively, according to a recent projection by the Energy Information Administration (EIA).[2] Another EIA estimate shows that the amount of tight oil and shale gas in the United States that could be extracted with today's technology would satisfy domestic oil consumption at current rates for approximately 8 years and domestic gas consumption for 25.[3]

How Will Shale Development Affect Energy Markets?

Total domestic production of oil and natural gas will continue to be higher than it would have been without shale development, reducing the prices of those energy supplies. The lower prices, in turn, will increase domestic consumption of oil and gas, domestic consumption of energy overall, and net exports of gas, while decreasing the production of oil and gas from conventional resources, net imports of oil, and the use of competing fuels.

Shale gas has affected energy prices in the United States more strongly than tight oil has, and it will continue to do so. Indeed, the Congressional Budget Office (CBO) estimates that if shale gas did not exist, the price of natural gas would be about 70 percent higher than currently projected by 2040—whereas if tight oil did not exist, the price of oil would be only about 5 percent higher. One reason for the difference is that shale gas is more plentiful than tight oil, relative to the size of their domestic markets. Another is that the North American market for natural gas is relatively insulated from conditions elsewhere by high transportation costs, so the effects of higher or lower domestic production on market prices are concentrated within the continent; oil, by contrast, is heavily traded in a worldwide market that diffuses the effects of domestic production on prices. (Oil prices are thus influenced by events that occur elsewhere in the world. For example, the recent sharp drop in crude oil prices—as of the end of November 2014, they had dropped about one-third from their recent peak in June—was caused not by any sudden or dramatic increase in the supply of tight oil during that period but by other factors, such as a rapid increase in Libyan production and a slowdown of consumption in Europe and Asia.)

EIA's projections of the development of shale resources are the most detailed currently available, and CBO considers them an appropriate basis for estimating the potential economic and budgetary effects of shale development. Nonetheless, like all projections of the future, they are subject to significant uncertainty. Many factors contribute to the uncertainty; for example, the abundance of shale resources, the fraction of those resources that will be recoverable with evolving technology, and the costs of recovering that fraction are not known for certain. Projections of more or less shale development would lead to larger or smaller estimates of the economic and budgetary effects.

How Will Shale Development Affect Economic Output?

The technological innovations behind hydraulic fracturing and horizontal drilling make existing labor and capital—whether they are employed in shale development, in industries using natural gas or oil, or in industries using products derived from natural gas or oil—more productive than they otherwise would be. That heightened productivity has increased gross domestic product (GDP) and will continue to do so.

Shale development also boosts GDP in other ways. The increase in GDP just described represents increased income, which allows people and firms to save and invest more in productive capital, and the higher productivity just described increases wages, raising the amount of labor available. Both the increased capital and the increased labor raise GDP. In addition, in the near term, shale development causes labor and capital to be used that would otherwise be idle, again raising GDP. In the longer term, however, whether shale resources are available or not, the labor and capital available in the economy will be used at roughly their maximum sustainable rates, so the additional labor and capital used to produce shale resources or energy-intensive goods will mostly be drawn away from the production of other goods and services. As a result, there will be no net change in GDP through that last route, although GDP will continue to be increased by shale development in the other ways just described.

On net, CBO estimates that real (inflation-adjusted) GDP will be about two-thirds of 1 percent higher in 2020 and about 1 percent higher in 2040 than it would have been without the development of shale resources. The actual effect on GDP could be higher or lower than that estimate, depending on the uncertain factors noted above—the abundance of shale resources, the fraction of those resources that will be recoverable, and the cost of developing that fraction—as well as on other considerations.

How Will Shale Development Affect the Federal Budget?

The increase in GDP resulting from shale development has increased federal tax revenues, and it will continue to do so. That increase will be slightly larger than the GDP increase in percentage terms, CBO expects. Specifically, CBO estimates that federal tax revenues will be about three-quarters of 1 percent (or about $35 billion) higher in 2020 and about 1 percent higher in 2040 than they would have been without shale development.

Shale production also contributes to federal receipts through payments that the developers of federally owned resources make to the government—but that contribution has been modest and will continue to be, because most shale resources are not on federal land. Working from EIA's projections of the future production of tight oil and shale gas, and also from its own forecasts of oil and natural gas prices, CBO estimates that federal royalties from shale (minus the amounts that the federal government transfers to the states) will be about $300 million annually by 2020.

What Policy Options Would Affect Shale Development?

There are a number of ways that the Congress could affect shale development and thus affect the oil and gas markets, economic output, and the federal budget. This report considers options that would change export policies—easing the current ban on exports of crude oil, repealing it, or changing the government's criteria for judging applications to export liquefied natural gas (LNG)—and concludes that the options would probably increase domestic production but have little effect on prices. That increase in production would probably make GDP and federal revenues slightly higher than they would be under current export policies.

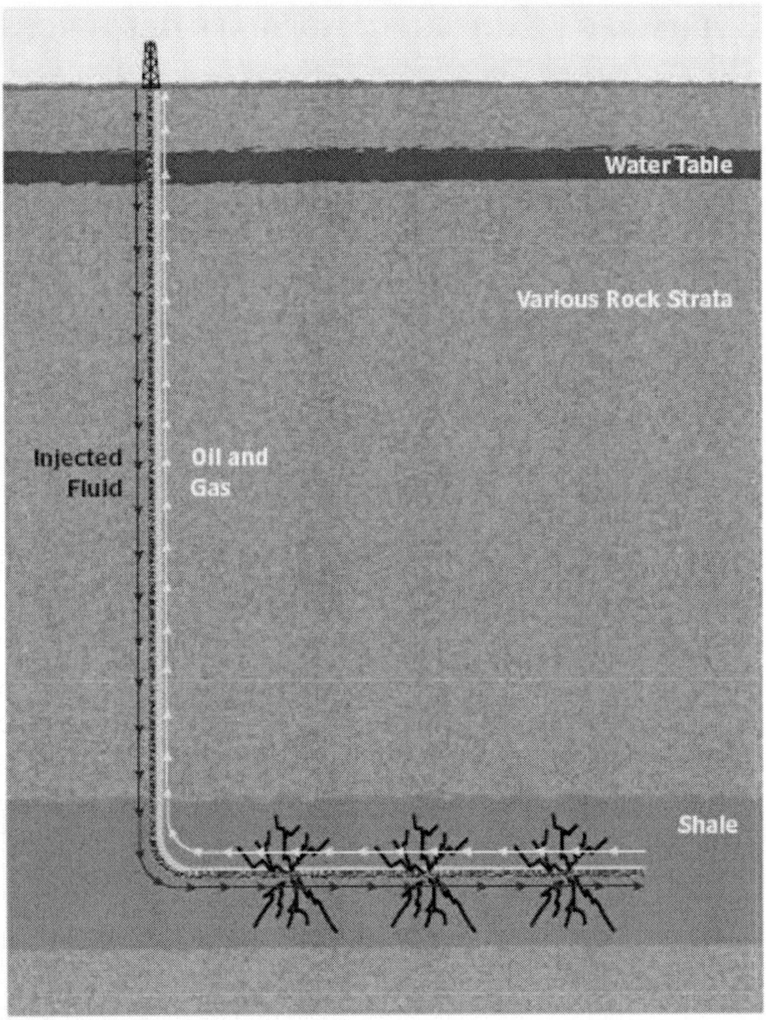

Source: Congressional Budget Office.

Figure 1. Hydraulic Fracturing and Horizontal Drilling.

Policy choices related to environmental regulation, such as whether the federal government should regulate further the environmental effects of shale development or leave such decisions to the states, are outlined in Appendix A. The Congress could also affect shale development through policies not considered here, such as those related to the infrastructure used to transport and process domestic shale gas and tight oil.

HYDRAULIC FRACTURING AND SHALE RESOURCES

Hydraulic fracturing, used with horizontal drilling and other advances in drilling technology, is a way to reach and extract natural gas and oil locked in certain rock formations, especially shale formations.[4] Some forms of hydraulic fracturing have been used to extract fossil fuels since the 1950s, but the method was not successfully combined with horizontal drilling for another 30 to 40 years, and it began to have a substantial impact on natural gas and oil production only in the past decade.

The process (often called fracing or fracking) begins with drilling a vertical well to the depth of a shale formation and, from there, drilling a horizontal well into the formation, which is much wider than it is thick (see Figure 1). A high-pressure mixture of water, chemicals, and small particles is pumped into the well to create fractures in the formation. Those fractures are held open by the particles as the injected fluid is withdrawn. Oil and gas then flow from the fractures into the well and up to the surface.

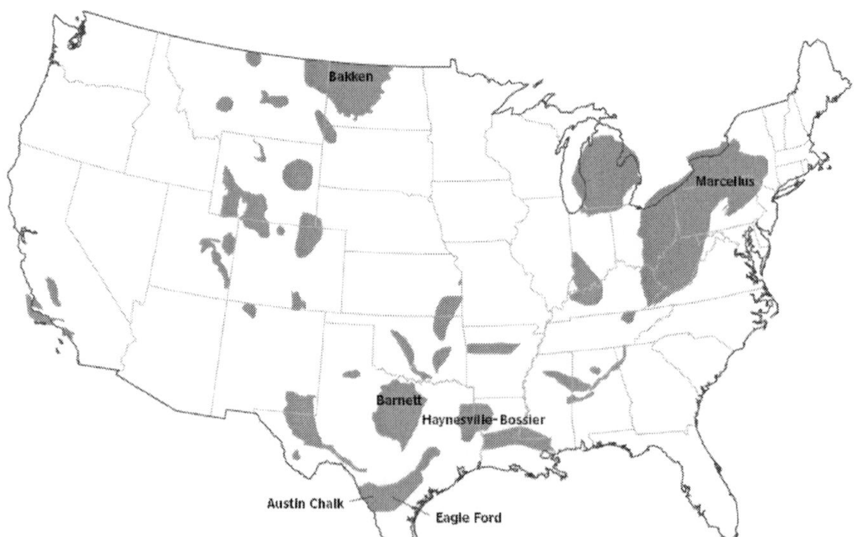

Source: Congressional Budget Office based on data from the Energy Information Administration, "Detailed Oil and Gas Field Maps" (accessed October 2, 2014), http://go.usa.gov/VKt4.

Figure 2. Shale Formations in the United States.

According to EIA estimates, of the shale gas in the United States that is technically recoverable—that is, that could be developed with current technology—25 percent is in the Marcellus Shale formation, which is located mainly in New York, Pennsylvania, and West Virginia (see Figure 2). The formations with the next-largest quantities of shale gas are the Haynesville-Bossier Shale in Texas and Louisiana, containing an estimated 15 percent of technically recoverable resources; the Eagle Ford Shale in Texas, containing about 10 percent of those resources; and the Barnett Shale, also in Texas, and also containing about 10 percent of those resources.

The Eagle Ford and Austin Chalk Shales in Texas, which are found at different depths but underlie some of the same land, together account for about 40 percent of technically recoverable tight oil. An additional 20 percent is in the Bakken Shale in North Dakota and Montana.[5]

EFFECTS ON ENERGY MARKETS

The production of shale gas and tight oil has risen dramatically over the past decade. The shale gas increase has been so large that, if it came from a separate country, that country would now be the world's third-largest natural gas supplier (behind first-place Russia and the U.S. supplies not from shale). Because of shale gas, domestic production of all natural gas is on pace to increase for the ninth straight year and has reached record highs (see Figure 3). With that increase in production, the wholesale price of natural gas in North America fell by about 70 percent, in inflation-adjusted terms, between 2008 and 2012, reaching its lowest level since 1998 (see Figure 4).[6] Though gas prices rebounded somewhat in 2013 and 2014, they remain low compared with the recent historical record.

The production of crude oil in the United States has also boomed, with domestic output up for the fifth straight year because of tight oil. But the increase in tight oil production, unlike the corresponding increase in shale gas, has had only a modest impact on prices.[7] As this report discusses below, the increase in tight oil is small relative to domestic oil consumption, and oil prices, unlike gas prices, are determined in a global market.

Shale development has affected other energy markets as well; for example, it has reduced the demand for coal. As the production of shale gas and tight oil increases, its effects on other energy markets—such as those for coal, for nuclear and renewable energy, and for energy-conserving equipment—will also increase. In addition, some analysts predict that energy-intensive production activities will increasingly relocate to the United States to take advantage of low gas prices.

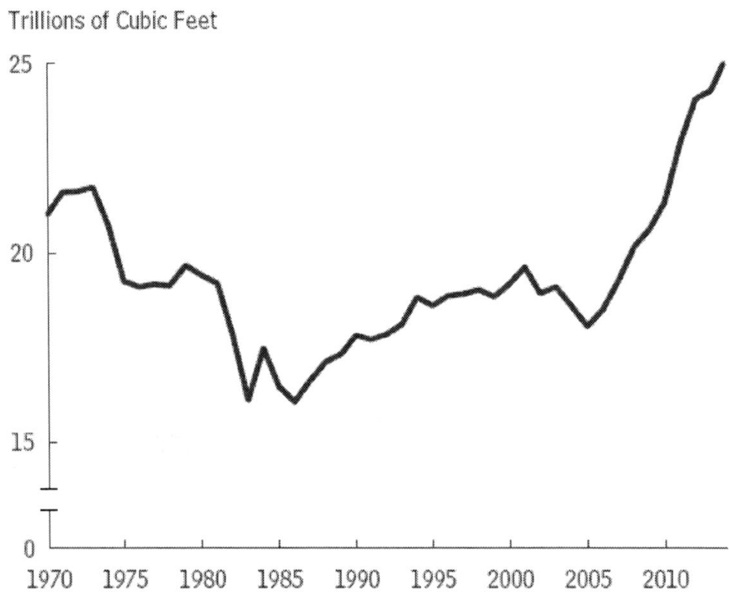

Source: Energy Information Administration, "U.S. Dry Natural Gas Production" (accessed November 14, 2014), http://go.usa.gov/7XgQ.

Note: Production in 2014 is an estimate based on monthly totals from January through September.

Figure 3. Annual Production of Natural Gas in the United States.

Trends in the Markets for Shale Gas and Tight Oil

According to EIA's projections, supply and demand conditions will keep the production of shale gas and tight oil growing in coming years—in fact, growing enough so that the overall domestic production of oil and gas will continue to grow, reversing the trend of the past several decades. Because of the growth in production, the domestic price of gas will be lower than it would have been in the absence of shale resources; so will the price of oil, though to a lesser extent. Those lower prices will boost oil and gas consumption and net exports of refined products and natural gas. (The quantity of exports will depend partly on federal policy choices, as the next section discusses.)

Production

EIA expects the production of shale gas to rise from 9.5 Tcf in 2013 to 20 Tcf in 2040. That increased production will be responsible for almost all the growth in overall U.S. gas supplies, which are projected to rise 13 Tcf over that time, from 24 Tcf in 2013 to 37 Tcf in 2040.

Also, EIA projects that the production of oil from shale formations will be 1.4 million barrels per day higher in 2020 than in 2013 but only 0.2 million barrels per day higher in 2040 than in 2013. That projection includes both tight oil and natural gas plant liquids, such as ethane, propane, and butane, which are sometimes obtained in the production of shale gas and are good substitutes for certain petroleum products. EIA expects that the production of tight oil alone will grow from 3.5 million barrels per day in 2013 to a peak of 4.8 million by about 2020; it will then fall back to 3.2 million by 2040, as production from existing wells wanes and new wells in less promising areas yield less oil. Natural gas plant liquids are projected to increase modestly but more steadily, from 2.5 million barrels per day in 2013 to 3 million in 2040.

Prices

The availability of shale energy (that is, shale gas and tight oil that come to the market) should lessen the growth of energy prices in the years ahead. Shale gas will probably have a larger impact in that way than tight oil will. One reason is simply that shale gas is more plentiful, relative to domestic consumption. Shale gas production today equals about 35 percent of total U.S. gas consumption, whereas tight oil production equals only about 15 percent of U.S. consumption of liquid fuels. By 2040, according to EIA's most recent long-term projection, shale gas will account for about 60 percent of all natural gas consumed in the United States, but tight oil will still represent only about 15 percent of all liquid fuels consumed.[8]

Another reason that U.S. natural gas markets will be more affected than domestic oil markets by shale energy is that they have far less international exposure. Natural gas markets are broadly split into three regions—North America, Europe, and Asia—and gas is transported within each of those regions by pipeline at relatively low cost. But the cost of transporting gas *between* regions is significant, primarily because it must undergo costly liquefaction before being shipped on oceangoing vessels. Therefore, little trade occurs between regions, and prices in the three markets are largely independent of one another. Crude oil, by contrast, can be moved around the world at relatively low cost by tanker ship or pipeline, which means that oil prices are approximately the same around the world. The effects of domestic shale gas production will therefore be concentrated in North America, whereas the effects of domestic tight oil production will be diffused internationally.[9]

Though shale energy is expected to lessen the growth of energy prices, continued growth in demand means that those prices will nevertheless continue to rise. EIA currently projects a doubling of the real price of natural gas in North America by 2040, as well as a roughly 30 percent increase in the real price of oil worldwide. Prices would be still higher if the production of U.S. shale energy turned out to be lower than EIA expects. For example, according to a recent EIA analysis, if shale resources turned out to be only half as abundant as the agency projected in its baseline scenario, domestic prices in 2040 would be about 40 percent higher for gas and somewhat less than 3 percent higher for oil than they would be under the baseline scenario.[10] (Similarly, CBO estimates that if shale resources did not exist at all, the price effects would be roughly twice as large, with gas and oil prices in 2040 that were roughly 70 percent and 5 percent higher, respectively, than currently projected.)

Consumption and Net Exports

Because of the lower prices that will result from shale development, the domestic consumption of gas will be higher than it would have been in the absence of shale resources; net exports of natural gas (that is, international consumption of domestically produced gas) will also be higher. Of the expected 13 Tcf increase in natural gas production between 2013 and 2040, EIA projects that 53 percent (about 7 Tcf) will be reflected in greater net exports and 47 percent (about 6 Tcf) in increased domestic consumption (see Figure 5). Roughly 75 percent of that projected increase in domestic consumption will be in the electric power and industrial sectors.

The electric power sector's increased consumption of gas will result not only from that sector's higher overall production of electricity but also from the growing importance of gas relative to other fuels. The EIA projection shows that by 2040, natural gas's share of the total fuel used in the electric power sector will grow from 21 percent to 25 percent; renewable fuels' share will also grow, while those of coal and nuclear power will shrink. In the industrial sector, by contrast, the increased use of natural gas is expected to be roughly in line with that sector's growth.

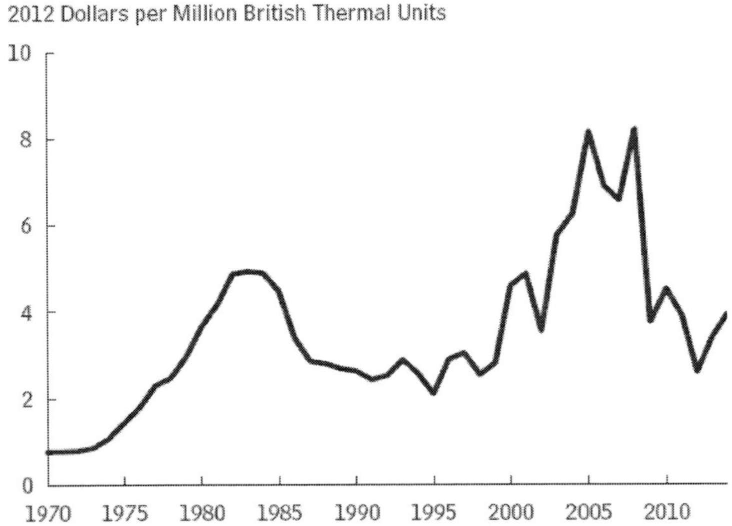

Source: Congressional Budget Office based on data from the Energy Information Administration, "U.S. Natural Gas Wellhead Price" (accessed November 4, 2014), http://go.usa.gov/7X2G, and "Henry Hub Natural Gas Spot Price" (accessed November 4, 2014), http://go.usa.gov/7X2z.

Note: Prices shown from 1970 through 2012 are average wellhead prices for gas produced nationally. Because the 2013 and 2014 averages are not available yet, averages for those years are derived from a historical relationship between average wellhead prices and the prices recorded at the Henry Hub natural gas pipeline interconnection in Louisiana. The Henry Hub price is commonly used as a benchmark for wholesale gas prices throughout North America. CBO converted prices into 2012 dollars by means of the GDP (gross domestic product) deflator.

Figure 4. Average Annual Price of Natural Gas.

Unlike the consumption of natural gas, the consumption of liquid fuel will be slightly lower in 2040 than in 2013, decreasing by about 1 percent, EIA estimates. The main reason is that changes in driving habits and improvements in vehicles' fuel economy are expected to reduce U.S. demand for liquid fuels. Also, the use of natural gas in the transportation sector is expected to grow, further lowering demand for petroleum.

The United States is currently a net importer of natural gas and of liquid fuels; that is, it consumes more than it produces (see Figure 6). But the production of shale resources has significantly reduced net imports—from nearly 4 Tcf of natural gas in 2007 to about 1.5 Tcf in 2013, and from 12.5 million barrels per day of liquid fuels in 2005 to 6.2 million in 2013. EIA projects that the United States will become a net exporter of natural gas in 2017 and remain so through 2040.[11] However, EIA expects the country to remain a net importer of liquid fuels throughout the projection period; net imports are projected to decline to about 5 million barrels per day as tight oil production increases, to stay steady for a few years, and then to return to current levels by 2040, as tight oil production falls.

Policy Options Related to Exports and Their Effects on Domestic Prices

To the extent that federal policy allows oil and gas to be imported and exported, their domestic prices reflect supply and demand not only in the United States but in other countries as well. Analysts and policymakers are currently proposing various changes to policies governing exports of crude oil and liquefied natural gas. In CBO's view, such changes would probably increase domestic oil and gas production, but they would probably have only a small effect on the domestic price of gas and a negligible effect on the domestic price of oil (which, again, is largely determined in the world market).

As this report discusses below, increases in oil and gas production resulting from shale development have boosted U.S. economic output and federal receipts and will continue to do so. The further increases in production that would result from the changes in export policies considered here would also have positive economic and budgetary effects, but smaller ones.

Exports of Crude Oil

Federal law prohibits the export of domestically produced crude oil, with few exceptions.[12] In 2013, only about 1 percent of crude oil produced in the United States (about 120,000 barrels per day) was exported, essentially all of it to Canada.[13] In contrast, federal policy does not restrict U.S. exports of refined petroleum products or of natural gas plant liquids, biofuels, and other nonpetroleum liquids. Together, exports of those fuels totaled 3.6 million barrels per day in 2013—a record high—with roughly half of that volume consisting of gasoline and diesel fuel.

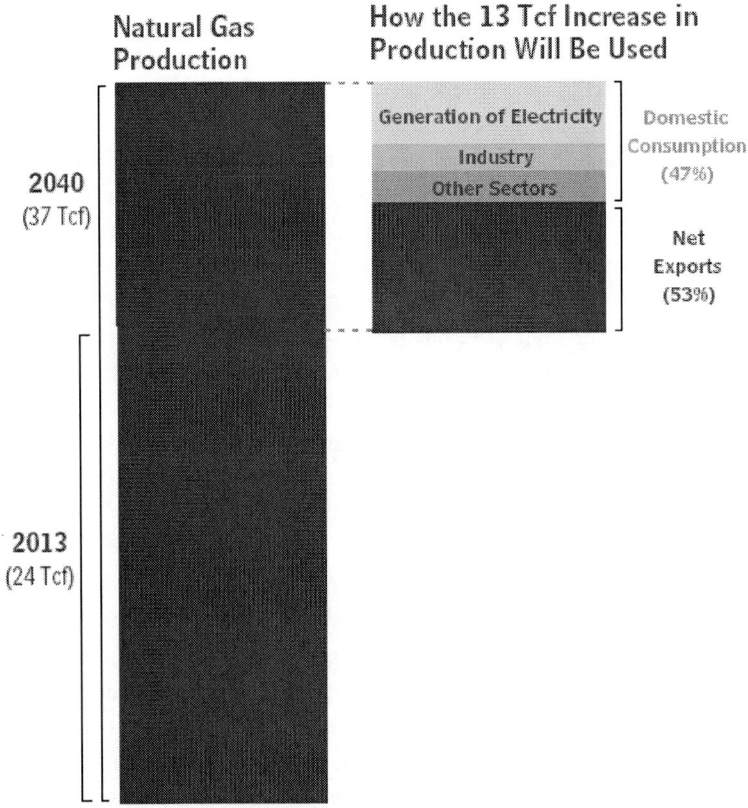

Source: Congressional Budget Office based on data from the Energy Information Administration, *Annual Energy Outlook 2014 With Projections to 2040*, DOE/EIA0383(2014) (April 2014), http://go.usa.gov/8KyF (PDF, 12 MB).

Note: Tcf = trillion cubic feet.

Figure 5. Natural Gas Production in the United States in 2013 and 2040.

Policy Options

Because U.S. supplies of crude oil have grown so dramatically in recent years, some policymakers have called for the ban on crude exports to be repealed. Current policy could also be changed less dramatically. For example, exports might be permitted not only to Canada but to certain other countries as well, such as Mexico, the countries of Central America, or all nations with which the United States had free-trade agreements (FTAs). Alternatively, the volume of allowed exports could be capped, or exports could be restricted to particular grades of crude oil.

Potential Effects of Those Options

Outright repeal of the ban on crude exports would probably lower world prices of oil and of liquid fuels produced from oil, but only slightly, and changes that left some export prohibitions in place would lower world prices even less. The reason is that prices depend on the total worldwide supply of crude oil, and the increase in total supply would probably be much smaller than the increase in the volume of U.S. crude exports. One recent study, for example, estimated that if the ban was repealed, U.S. crude exports would increase by as much as 1.5 million barrels per day, but world supply would increase by no more than 200,000 barrels per day—less than one-quarter of one percent of the current total.[14] Two factors explain the difference. First, what contributes to the total worldwide supply of oil is not U.S. *exports* but U.S. *production*, which would rise much less than exports would. For instance, the study estimated that if the ban was repealed and crude exports rose by about 1.5 million barrels per day, U.S. oil production would rise by only about 500,000 barrels per day.[15] (Domestic consumption would not change much, however, because U.S. crude *imports* would be higher as well, as the next paragraph explains.) Second, the net increase in world production would be smaller even than the increase in U.S. production, because the U.S. increase would drive some competing high-cost supplies from the market.

Perhaps counterintuitively, U.S. consumers of gasoline, diesel fuel, and other oil products would probably benefit, along with domestic oil producers, if the ban was repealed; domestic refiners would be adversely affected, as would foreign oil producers. Consumers would benefit from small reductions—5 to 10 cents per gallon, in the baseline scenario of a recent study—in the domestic prices of oil products, because those prices depend primarily on the world price of crude oil, which would decline slightly once lower-priced U.S. crudes were available in the international market.[16] By contrast, the prices of domestic light crude oils (which include tight oils) seen by some U.S. crude oil producers and petroleum refiners would rise. Refineries in the United States are better configured than refineries abroad to process heavy crudes, so under the current ban, light crudes are less valuable—and therefore sell for less—in the United States than in the global market.[17] If the ban was repealed, some domestic refiners would continue to buy light crudes, and others would increase their imports of heavy crudes; in either case, the cost of their crude oil inputs would be higher than it had been under the ban, and because they would continue to sell their refined products at levels closely linked to the world price of oil, their profits would fall.

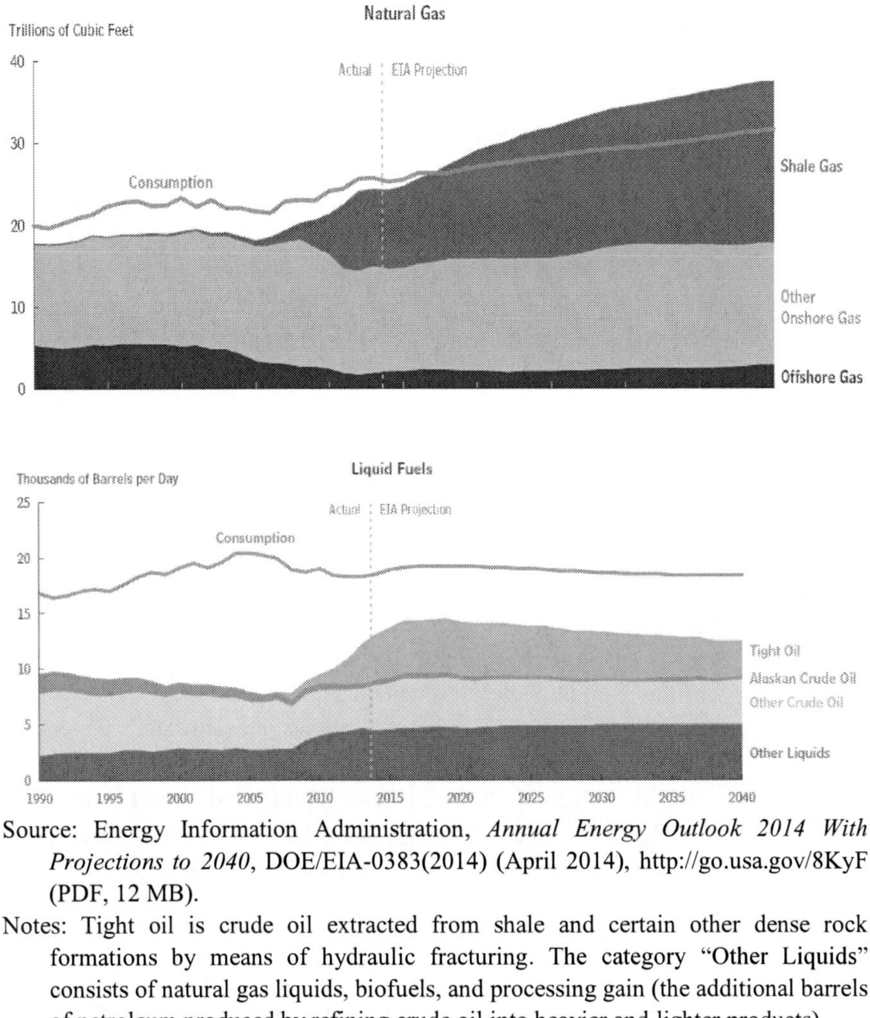

Source: Energy Information Administration, *Annual Energy Outlook 2014 With Projections to 2040*, DOE/EIA-0383(2014) (April 2014), http://go.usa.gov/8KyF (PDF, 12 MB).

Notes: Tight oil is crude oil extracted from shale and certain other dense rock formations by means of hydraulic fracturing. The category "Other Liquids" consists of natural gas liquids, biofuels, and processing gain (the additional barrels of petroleum produced by refining crude oil into heavier and lighter products).

EIA = Energy Information Administration.

Figure 6. Consumption and Supply of Natural Gas and Liquid Fuels in the United States.

Exports of Natural Gas

The United States trades significant quantities of natural gas with Canada and Mexico. Using pipelines, it currently exports about 1.6 Tcf per year to those two countries, and it imports about 2.9 Tcf per year, virtually all of it from Canada.[18] The pipelines through which gas travels between Canada, the

United States, and Mexico create a unified North American market in which the price of gas is determined by the total supply and demand of all three countries. Once the United States is a net pipeline exporter, as EIA projects it will be within a decade, domestic gas prices will be higher than they would be without pipeline exports.[19]

The only way to transport significant volumes of natural gas to countries that are not connected to the United States by gas pipelines is to liquefy the gas and move it by ship. The United States has very little capacity to do that, because it was expected until recently to be a substantial importer for decades to come.[20] But as hydraulic fracturing and related technologies have become widespread in the United States, natural gas has become much cheaper here than in foreign markets; in 2013, average gas prices were about three times higher in Europe and about four times higher in Japan, both of which are large gas consumers.[21] Such price differences, if they last, could make selling LNG overseas profitable, despite the significant cost of liquefying natural gas, transporting it, and converting it back into gaseous form.

Policy Options

Restrictions on gas trade by pipeline are not allowed under the free-trade agreements that the United States has with Canada and Mexico. Exports of LNG, however, are subject to restrictions under current law, which the Congress could modify in various ways.

Currently, the construction of facilities to liquefy and export natural gas requires approval from the Federal Energy Regulatory Commission (FERC), and the exports themselves must be approved by the Department of Energy (DOE). Prospective exporters can apply for blanket authority to ship LNG to countries in either or both of two groups: those with FTAs with the United States that cover natural gas and those without such agreements.

Exports to countries without FTAs, which account for roughly 80 percent of LNG imports worldwide, require DOE to determine that the exports would be in the public interest; DOE regards LNG exports to FTA countries as automatically being in the public interest.

As of October 2014, four LNG export terminals, proposed for Louisiana, Maryland, and Texas, had received approval for construction and for exports to countries without FTAs; if those terminals are built in the next several years, their combined capacity of about 2.5 Tcf will represent roughly 8 percent of North American gas consumption. Four more facilities, proposed for Florida, Louisiana, and Oregon and having a combined capacity of 1.5 Tcf, have recently been authorized by DOE to export LNG to countries without

FTAs, but none of them have received approval for construction from FERC. All told, the roughly 30 applications that have sought full approval from FERC and DOE would create facilities that could export about 13 Tcf of LNG per year, an amount equal to roughly 40 percent of the natural gas consumed in North America.[22]

If the Congress wanted to change LNG export capacity, it could alter the criteria for DOE's approval of such exports to countries without FTAs. For example, it could require DOE to treat applications to export to those countries the same way that it treats applications to export to FTA countries—that is, automatically assuming that they are in the public interest. Such a change would speed the review process and make approvals more likely. Alternatively, the Congress could change the federal review process to make approvals of LNG exports less likely. For instance, when determining whether to allow LNG exports to a country without an FTA with the United States that covered natural gas, DOE could be required to give particular weight to the effects that the resulting higher domestic gas prices would have on low-income households in this country.

Potential Effects of Those Options

If the pending applications were approved, and export capacity of 13 Tcf per year was built and fully used, that 40 percent decline in supply in North America would boost gas prices considerably (unless suppliers greatly increased production in response to even a small price increase). But that much capacity might be approved without being built or fully used, in which case the actual volume of LNG exports might be much smaller. Whether approval of facilities led to construction and use would depend heavily on whether the difference between North American and overseas gas prices remained large enough to justify the costs of producing and exporting LNG.

On the one hand, today's large price gap could narrow— for example, if some new LNG facilities began operating, increasing the supply of natural gas overseas and reducing its price there while raising it here. The price gap could also narrow if major foreign suppliers of natural gas increased production to protect their market share; if new overseas gas resources, particularly shale gas, came to market; or if North American demand grew faster than supply. On the other hand, the price gap could widen further in the future, giving domestic firms an even bigger incentive to export gas—if, for example, worldwide demand for gas remained high but little additional LNG liquefaction and export capacity was built outside North America, or if North American gas supplies grew faster than demand.

A 2012 study commissioned by DOE analyzed future gas prices in various scenarios with different supply and demand conditions and different amounts of available export capacity. Most of the scenarios showed a future gap between U.S. and international gas prices too small to create much overseas demand for U.S. LNG; exports in those scenarios were accordingly small or nonexistent, even if a large amount of export capacity was approved.[23] Those findings were broadly confirmed in a 2014 update of the study (not commissioned by DOE), which found that under expected supply and demand conditions, allowing 2 Tcf or more of export capacity would result in only a 2 percent to 5 percent increase in domestic gas prices.[24] However, in scenarios in which sizable export capacity was approved and fully used, domestic prices would rise more sharply. For example, the more recent study estimated that with exports of about 4 Tcf per year—an amount consistent with the capacity of projects already approved by DOE, including those awaiting approval from FERC—domestic natural gas prices would probably be about 15 percent higher than they would have been with no export capacity. It also estimated that with exports of about 13 Tcf per year—roughly the capacity of all LNG facilities that have been approved or are currently seeking approval—prices would be 30 percent to 45 percent higher. But to support such high exports, overseas demand for LNG from the United States or U.S. supplies of shale resources (or both) would have to be much higher than projected by EIA at the time the studies were conducted.

EIA's most recent projection, which CBO uses in its baseline scenario, has the United States exporting about 2 Tcf per year of LNG by 2020 and about 3.5 Tcf by 2040.[25] Taking into account differences in the economic conditions underlying EIA's projection and those underlying the studies discussed above, CBO estimates that the projected 3.5 Tcf of exports would increase domestic gas prices in 2040 by 10 percent or less, relative to the prices that would exist with no exports of LNG.[26]

To the extent that market conditions supported LNG exports, making capacity available to allow those exports would raise GDP—in part because more domestic gas would be produced, but also because the gas would be sold overseas at higher prices than at home. However, that increased GDP would not accrue to people in the United States uniformly. Higher prices for gas exported overseas would mean greater profits for U.S. gas producers; but the fact that domestic prices, too, would rise would mean that U.S. gas consumers faced higher costs. One of the studies mentioned above estimated that an increase in North American gas prices of $1 per million British thermal units, or Btus (about five times the increase found in the baseline scenario of the

2014 update to the DOE study), would increase costs for gas and electricity by $50 per year for U.S. households with annual income less than $20,000 and by $90 per year for those with annual income above $100,000.[27] Some households—for example, those that owned shares of companies that produced gas, those that owned land in gas-rich areas, and those with members employed in the gas industry—would enjoy higher income that at least partly offset, if not outweighed, the increased gas and electricity costs.

Another effect of LNG exports would be to increase the integration of the North American gas market with the European and Asian markets. That would both increase the exposure of domestic consumers to supply shocks overseas and ameliorate the domestic effects of reductions in North American supplies. However, a full consideration of the effects of LNG exports on household income and market integration is beyond the scope of this report.[28]

Uncertainty in the Projections

Projections of shale development's impact on energy markets are inherently uncertain. A recent illustration of the uncertainty was EIA's energy market forecast in 2012, which projected that 2013 tight oil and shale gas production would total 0.9 million barrels per day and 7.6 Tcf, respectively.[29] The agency now expects 2013 production to have totaled 3.5 million barrels per day and 9.4 Tcf.[30]

There are many reasons for the uncertainty. Some involve the future availability of shale resources and others the future demand for those resources. Uncertainty also exists about the factors that have affected recent gas prices, and those factors influence estimates of future prices.

The Availability of Shale Resources

The main reason for the difficulty of projecting the market effects of shale development, CBO believes, is uncertainty about the future availability of shale resources. To estimate future availability, analysts must assess three items, each of which is itself uncertain:

- The total quantity of shale resources in the ground;
- The quantity of technically recoverable resources at various points in the future, which is the fraction of total resources that could be recovered at each of those points with the technology then available, and

- The costs of developing those technically recoverable resources (which are relevant to future availability because developers would not extract resources that could be developed only at an exorbitant cost).

What makes the first two items uncertain is clear: Not all shale resources have been identified, and improvements in technology are difficult to predict. In 2011, for example, the U.S. Geological Survey (USGS) released an estimate of technically recoverable shale gas from the Marcellus Shale that, because hydraulic fracturing and horizontal drilling had made more resources recoverable, was about 40 times larger than its previous estimate, which was issued in 2003.[31] However, in its annual estimates, EIA had expected even greater growth, so it lowered its estimate of the shale gas present in the Marcellus by about 65 percent after considering the USGS finding.[32] Similarly, on the basis of production trends and a revised understanding of the area's geology, EIA recently lowered its estimate of technically recoverable tight oil reserves in California's Monterey Shale formation by 96 percent.[33]

The third item, the cost of developing technically recoverable resources, is uncertain for many reasons. One is that limited information exists about the rate at which tight oil and shale gas wells become less productive over time. Initial evidence suggests that production declines much more rapidly for tight oil and shale gas wells than for conventional oil and gas wells; some estimates suggest that it falls 80 percent or more over the first three years of operation. However, because shale wells have not been in operation very long, it is difficult to draw firm conclusions about their lifetime rates of production. If the productivity of tight oil and shale gas wells turns out to decline more slowly than experts project, current estimates will have overstated the cost of producing a given quantity of shale energy from a given quantity of resources. The reverse will be the case if productivity declines more quickly than expected.

Another reason that production costs are uncertain is that limited information is available about the distribution of well productivity. A recent analysis found that the amount of shale gas that new wells in the same rock formation yielded in the first few months was distributed very unevenly, with high production from relatively few wells and low production from the rest.[34] As production diminishes in the areas that were developed first because they were considered the most promising, and as development moves into other areas, the percentage of wells that are highly productive may fall, raising the cost of finding such wells and thus the average cost of developing shale resources. Alternatively, if exploration methods improve, the percentage of wells that are highly productive may increase.

The Demand for Shale Energy

Domestic and foreign demand for shale energy depends on many uncertain factors, such as population growth, economic growth, the cost of conventional oil and gas, the cost of other competing energy sources, and the energy intensity of the economy (that is, the average amount of energy used in producing a dollar's worth of GDP).[35] None of those factors can be forecast with precision. For instance, driving habits in the future, the supply of conventional oil and gas, and the cost of generating electricity from renewable fuels might differ from what is currently expected. Projections of foreign demand are subject to additional uncertainty about transportation costs, other costs of trade, and foreign governments' trade policies.

The Factors Influencing Recent Gas Prices

Still another source of uncertainty in estimates of the effects of shale development involves the extent to which the recent declines in gas prices reflect factors other than the increased availability of shale gas. At least three such factors may be at work. First, the recent recession and slow economic recovery have reduced gas prices by reducing demand. Second, producers may be supplying more gas than they would have otherwise, given current prices, because of the boom in gas development that occurred in the second half of the last decade, when gas prices were much higher. Third, gas production is being supported by oil prices, which have been fairly high until recently. Gas producers in areas rich in natural gas plant liquids— which, as this report noted earlier, are sometimes obtained in the production of shale gas and are good substitutes for certain petroleum products—have been willing to produce and sell natural gas at a loss so that they can obtain those liquids, the prices of which are more closely linked to the price of crude oil than to the price of natural gas.[36]

The more those three factors (or others) have been responsible for today's low gas prices, the smaller a role has been played by the current availability of shale gas, and the more estimates may overstate the future sensitivity of gas prices to the availability of shale gas.

Suppose, for instance, that more of the decline in gas prices since 2008 resulted from the economic slowdown than a particular model accounts for, and that less resulted from the growth of shale gas availability. In that case, the model would have overestimated the past effects of shale gas availability on prices, and projections based on that model would similarly overestimate the future effects of shale gas availability on prices.

EFFECTS ON ECONOMIC OUTPUT

In the long run, CBO estimates, the development of shale resources will lead to higher GDP by increasing the productivity of existing labor and capital and by increasing the amount of labor and capital in use. Specifically, CBO projects that real GDP will be 0.7 percent higher in 2020 and 0.9 percent higher in 2040 than it would have been without shale development.

In recent years, shale development has probably had a larger effect on GDP, having employed labor and capital that would otherwise have been unused because of weak demand for goods and services. That larger effect will probably persist over the next few years—that is, as long as interest rates remain low and output remains distinctly below its maximum sustainable level. But after output moves back toward its maximum sustainable level, labor and capital used to produce shale resources or gas-intensive goods will mostly be drawn away from the production of other goods and services, which means that shale development will have a smaller net effect on GDP.

In the Next Few Years

Shale development has boosted GDP in recent years and will continue to do so. However, CBO has not quantified the effect over the next few years, because shale development's short-term effects on the economy, other than on the output of oil and gas, are especially difficult to measure. Those effects include increased investment in the oil and gas industry and in industries that support it; increased investment and production in other industries because energy prices are lower than they would otherwise be; and increased demand for goods and services because of greater household income—all of which increase GDP. Shale development also reduces the amount of labor and capital available for other uses and reduces the production of energy from conventional resources; those effects reduce GDP.

Increased Output of Oil and Gas

Shale development has increased U.S. output of tight oil and shale gas, raising GDP. The market value of shale gas produced in 2013 (reflecting the contributions of both the gas industry and the other industries that supply goods and services used to produce shale gas) was about $35 billion. In the same year, the market value of tight oil, including natural gas plant liquids produced by hydraulic fracturing, was about $160 billion. Combined, sales of shale gas and tight oil therefore totaled about $195 billion, or roughly 1.2 percent of GDP.

Increased Investment in the Oil and Gas Industry and in Supporting Industries

Shale development has probably raised GDP in recent years through greater spending on the development of new wells. Between 2004 and 2012, investment in the oil and gas extraction industry increased from 0.4 percent of GDP to 0.9 percent.[37] However, that increase included investment in conventional oil production that probably would have occurred even without shale development because of the sharp rise in oil prices over that period. CBO did not estimate how much of the increase in investment could be attributed to shale development.

In addition, industries that support the oil and gas sector have spent more on new facilities and equipment, such as pipelines and trains, as a result of shale development. CBO did not quantify that increase in investment either.

Increased Investment and Production in Other Industries

Industries that use natural gas intensively— such as the steel, petrochemical, fertilizer, and electricity industries—have expanded production to take advantage of energy prices that are lower than they would have been without shale development. Such industries have become more competitive internationally because of the fall in energy prices in the United States: A number of new U.S. petrochemical and fertilizer facilities are being planned, for example, and one company is in the process of moving two methanol plants from Chile to Louisiana.[38] Thus, shale development has boosted GDP by raising investment in, and production from, energy-intensive industries—but it is very difficult to estimate the magnitude of that effect.[39]

Increased Demand

Higher employment resulting from shale development, along with a larger capital stock resulting from increased investment in the development and use of shale resources, has led to higher household income and thus greater demand for goods and services. Some of that increased demand has been met by the additional production from the energy-intensive industries described above. However, much of the increase has been for products supplied by firms that do not directly benefit from lower natural gas and oil prices. In order to meet the increased demand, those firms have increased employment and investment, which has raised GDP still further in the short term.

Less Labor and Capital Available for Other Uses

The effects described above have shifted some workers and capital away from other uses, which means that some economic activity has been forgone. That forgone output has partly offset shale development's positive effects on GDP. Although CBO has not quantified the forgone output, the fact that the economy's slow recovery from the recent recession has left many resources underused suggests that the amount is small.

Less Production from and Investment in Conventional Energy Resources

As shale development has made energy prices lower than they would have been otherwise, the production of gas and oil from some conventional supplies has become unprofitable and has therefore been abandoned, and some investment in conventional sources of gas and oil has not been undertaken. Similarly, electric utilities' substitution of natural gas for coal has reduced production from and investment in coal mining. The forgone production and investment, like the reduced output from sectors that lost labor and capital, has partly offset shale development's positive effects on GDP. CBO estimates that conventional gas production falls by about one-tenth the amount of an increase in shale gas production, which means that the resulting loss in GDP in 2013 was probably small.[40] The effect of reduced investment in conventional energy on GDP in 2013 is harder to quantify.

In the Longer Term

Shale development will raise GDP in the longer term in two ways: increasing the productivity of existing labor and capital, and increasing the amount of labor and capital in use. CBO estimates that, as a result, real GDP will be 0.7 percent higher in 2020 and 0.9 percent higher in 2040 than it would have been without shale development, although those estimates are subject to considerable uncertainty.[41] The longer-term effects of shale development on GDP will probably be smaller than the near-term effects described above (see Box 1).

Increased Productivity

Shale development raises GDP by increasing the productivity of labor and capital. That increased productivity is projected to make GDP 0.4 percent higher in 2020 and 0.5 percent higher in 2040 than it would have been in the absence of shale development.

Some of the increased productivity comes from the labor and capital used in shale development itself, which are more productive because of hydraulic fracturing and horizontal drilling than they would have been without those techniques. CBO estimates that the value of the tight oil and shale gas produced in both 2020 and 2040 will be about 1.3 percent of real GDP. But in the absence of hydraulic fracturing and horizontal drilling, CBO estimates, the labor and capital now projected to be used to produce that output would contribute only about 1.0 percent to GDP in 2020 and about 0.9 percent in 2040. The boost to GDP from reallocating labor and capital into the production of tight oil and shale gas is the difference between those estimates: about 0.3 percent of GDP in 2020 and 0.4 percent in 2040. (For details about that estimate and others in this section of the report, see Appendix B.)

Another component of the increased productivity resulting from shale development comes from replacing high-cost conventional oil and gas with shale resources. Because less labor and capital are now required to produce the same amount of oil and gas, the shift frees up labor and capital, which are used to produce other goods, thereby increasing GDP. However, because the reduction in conventional oil and gas will be modest, the resulting increase in GDP will be small in both 2020 and 2040.

The rest of the increased productivity comes from labor and capital that are used more efficiently elsewhere in the economy because of increased consumption of oil and gas. As energy-intensive products and methods of production grow cheaper, the same amount of output can be produced with less labor and capital. For example, as the cost of generating electricity from gas has fallen, some electric utilities have increased their productivity by switching from coal to gas. Through such shifts, GDP will be about 0.1 percent higher in both 2020 and 2040 than it would have been without shale development, CBO estimates.

Higher output would also result if shale development led manufacturing to become a larger share of the economy and if labor was generally more productive in manufacturing than in other sectors. However, recent earnings data do not demonstrate that labor productivity is higher in manufacturing. Although the average weekly earnings of employees in manufacturing were higher than those of all private-sector employees in 2013, hourly earnings were about the same, meaning that most of the difference in weekly earnings was due to a longer average workweek in manufacturing.

Box 1. Why the Economic Effects of Shale Development Will Be Larger in the Near Term

The positive effects of shale development on gross domestic product (GDP) are partly offset by output that is forgone when labor and capital are shifted away from other uses. That offsetting effect has not been large so far, in the Congressional Budget Office's assessment, because the economy's slow recovery from the recent recession has left many resources unused. However, the effect will be larger once the economy moves back toward producing its maximum sustainable level of output. At that point, the labor and capital shifted into the production of shale resources or energy-intensive goods and services will mostly be drawn away from the production of other goods and services. Consequently, shale development's net effect on GDP is likely to be smaller in the longer term than in the near term.

The redistribution of labor and capital will occur in various ways. For example, some higher-cost production of natural gas from conventional resources will become unprofitable, pushing labor and capital elsewhere. The composition of domestic production will shift toward energy-intensive manufacturing and away from other industries. And increased net exports of natural gas and oil will boost the value of the dollar, making goods produced in the United States more expensive relative to U.S. imports and therefore leading to reduced production of those goods. (Economists refer to that phenomenon as Dutch disease, remembering the discovery in 1959 of the Groningen gas field in the Netherlands, which led to large exports of natural gas and a surge in the value of the Dutch currency in the late 1960s and early 1970s—and thereby made Dutch manufacturing less competitive.) The increase in the value of the dollar will probably be small, but it will affect all U.S. exports and imports and would probably have a discernible effect on the economy.

A recent study illustrates the difference between shale development's effects on GDP in the near term and in the longer term. An average of the conservative and optimistic scenarios in the study indicates that shale resources are expected to boost maximum sustainable GDP by 0.65 percent and actual GDP by 1.35 percent between 2013 and 2020.

The difference between those estimates illustrates the additional response of GDP to shale development when the economy is not operating at its maximum sustainable level of output.[1]

[1] See Trevor Houser and Shashank Mohan, *Fueling Up: The Economic Implications of America's Oil and Gas Boom* (Peterson Institute for International Economics, 2014), Chapter 4, http://bookstore.piie.com/book-store/6567.html.

Increased Total Labor and Capital

Shale development will also raise GDP by increasing the amounts of labor and capital used in the economy, in CBO's assessment. That increase will happen in at least two ways. First, the increase in output generated by higher productivity that was described above will result in additional income; part of that income will be saved and then invested, increasing the capital stock. Second, the higher productivity will increase wages, improving the return to workers from each hour of work and encouraging them to work more. Because of those effects, CBO estimates, GDP will be 0.3 percent higher in 2020 and about 0.4 percent higher in 2040 than it would have been without shale development.[42]

Other effects of shale development on the total amounts of labor and capital (and in turn on GDP) are highly uncertain, so CBO did not estimate them. For example, if the industries that produce and use natural gas and oil, or those that supply infrastructure for shale development, are more capital-intensive than those that see production fall as a result of shale development, the demand for capital and thus the overall return on investment in the United States will be higher. That higher rate of return will lead to increases in private saving and in capital inflows from abroad. But under the same circumstances, companies' desire to replace labor with capital will reduce the return to working, reducing the labor supply.

Other Considerations

Two more considerations should be mentioned that are related to shale development's effects on the economy in the longer term. One involves a reduction in the dollar cost of U.S. imports; the other involves uncertainty in the estimates of economic effects.

The Cost of Imports

Shale development confers an economic benefit that raises the standard of living in the United States but does not show up as greater GDP. Specifically, increased net exports of natural gas and oil boost the value of the dollar, making imports cheaper and allowing consumers to buy more and businesses to invest more for a given quantity of exports and a given amount of GDP. CBO has not quantified that effect, however.

Uncertainty

CBO's estimates of shale development's effects on GDP are accompanied by significant uncertainty of various kinds. The estimates rest on baseline projections of the prices of shale gas and tight oil, of the quantities of those fuels produced in the United States, and of the profitability of that production—and as is explained earlier (in the section "Uncertainty in the Projections"), all of those projections are uncertain, because of underlying uncertainty about demand for natural gas and oil, demand for other forms of energy, the availability of shale resources, and exploration and production technology.

CBO therefore estimated the effects of shale development not only according to those baseline projections but also under two alternative scenarios. In the first scenario, prices, production, and profitability are all lower than projected in the baseline. Prices of natural gas and oil (reflecting recent EIA projections of price uncertainty) are about 25 percent lower in 2015 than they are in the baseline projection, then grow more slowly than they do in the baseline, and are about 50 percent lower by 2040. The production of shale gas and tight oil is about 40 percent lower than in the baseline by 2040, a figure that is consistent with what EIA calls its low-resource scenario. And the average cost of producing shale gas rises 75 percent as quickly as the price of natural gas, compared with 50 percent as quickly in CBO's baseline projection.[43]

In the second alternative scenario, the three factors are all *higher* than projected in the baseline. The prices of natural gas and oil start out about 35 percent higher than they are in CBO's baseline projection and grow to be roughly 50 percent higher.[44] The production of shale gas and tight oil is about 40 percent higher than in the baseline by 2040; and profitability is higher because the average cost of producing shale gas rises only 25 percent as fast as the price of natural gas.

In the first scenario, shale development makes real GDP 0.4 percent higher in 2020 and 0.3 percent higher in 2040 than it would have been otherwise. (The effect is smaller in 2040 because the economy then will be larger relative to the market value of shale energy in the scenario.) In the second scenario, GDP is 1.3 percent higher in 2020 and nearly 2 percent higher in 2040 because of shale development.

EFFECTS ON THE FEDERAL BUDGET

The development of shale resources affects two kinds of federal receipts. The first, federal tax revenues, rise as shale development boosts GDP. The second, payments to the government by private developers of federally owned resources, also increase with shale development—but not much, because most of the nation's shale gas and tight oil is not owned by the federal government.

Tax Revenues

The development of shale resources has increased economic activity in recent years and will continue to do so, as the previous section explains. That increased activity is reflected in higher income of various kinds, such as wages and salaries, income from partnerships and sole proprietorships, dividends, and corporate profits. And because that higher income is subject to a combination of individual income taxes, corporate income taxes, and payroll taxes, shale development increases federal tax revenues as well.

CBO expects the effect of shale development on revenues to be slightly higher in percentage terms than the effect on GDP. As the previous section also mentions, CBO estimates that real GDP will be 0.7 percent higher in 2020 and 0.9 percent higher in 2040 than it would have been without shale development. On the basis of that increase in GDP, CBO estimates that revenues will be higher by 0.8 percent (or about $35 billion) in 2020 and by 1.0 percent in 2040 than they would have been without shale development.[45]

In arriving at those rough estimates for 2020 and 2040, CBO assumed that the net effect of shale development on GDP would be allocated among the various types of taxable and nontaxable income, and across households in different tax brackets, in the same proportions in which overall GDP was expected to be allocated. Because the United States has a progressive individual income tax system—that is, one in which income in higher brackets is taxed at higher rates—that assumption led to the conclusion that shale development would have a greater effect on revenues than on GDP, in percentage terms. (By contrast, if the GDP added by shale development was unusually concentrated among people in the lowest tax bracket, shale development might have a smaller percentage impact on revenues than on GDP.)

Payments for Federally Owned Resources

The federal government receives payments from private developers of federally owned oil and gas. In the case of onshore oil and gas, about 90 percent of the payments are royalties on production; the rest are payments to obtain leases and rent on leases not yet put into production. All of the payments go initially to the U.S. Treasury, but under current law, the federal government generally pays about half to the states from which the resources were extracted. After adjusting for those payments to states, CBO estimates that net federal royalties from all onshore oil and gas leases will average about $1.4 billion a year over the 2015–2024 period.[46]

The portion of those royalties attributable to tight oil and shale gas production is uncertain because the government does not supply data breaking down production from federal lands by geologic formation. However, on the basis of information from state agencies and EIA, CBO estimates that in 2012, shale gas probably accounted for about 3 percent of onshore natural gas produced from federal land and tight oil for about 25 percent of onshore oil produced from federal land. Those estimates accord with the observation that few of the country's current and potential sources of shale gas and tight oil lie beneath federally owned land (see Figure 7).[47] For example, the Rocky Mountain region, which accounts for almost all of the natural gas (including shale gas) produced on federal land, is a minor source of the nation's supply of shale gas, yielding less than 1 percent of the total through 2024, according to EIA's projections.

In addition, the production of shale energy reduces the value of natural gas produced from federal *offshore* leases. That reduction diminishes the royalties (which are based on sales value) paid by the developers of those resources, and it therefore also diminishes the net effects of shale development on federal receipts.

On the basis of the preceding estimates and CBO's current forecast of oil and gas prices, along with EIA's projections of domestic production, CBO estimates that net federal royalties from tight oil and shale gas will total about $300 million a year by 2024. CBO anticipates that most of those royalties will come from the production of tight oil in the southeastern corner of New Mexico and in the Rocky Mountain region.[48] As with other estimates in this report, the $300 million figure is subject to substantial uncertainty.[49]

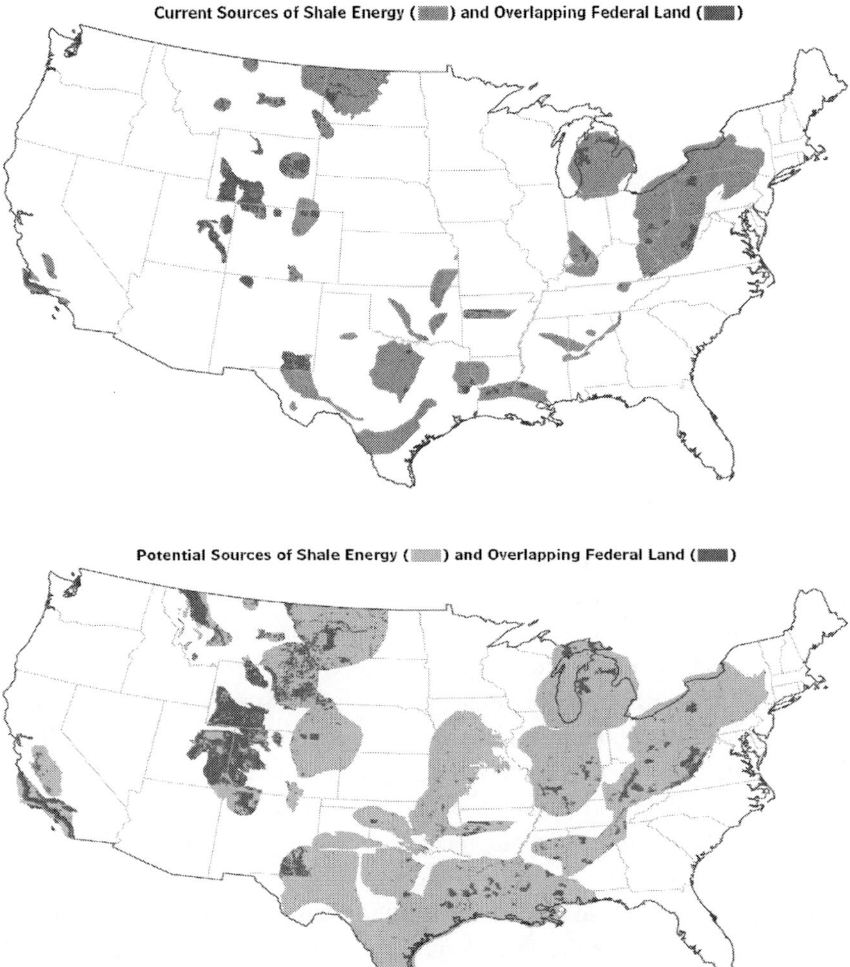

Source: Congressional Budget Office based on data from the Energy Information Administration, "Detailed Oil and Gas Field Maps" (accessed October 2, 2014), http://go.usa.gov/VKt4, and from the U.S. Geological Survey, "Federal Lands of the United States" (accessed October 2, 2014), http://go.usa.gov/vdzT.

Note: Shale energy is oil and natural gas extracted from shale and certain other dense rock formations by means of hydraulic fracturing.

Figure 7. Shale Formations and Federal Land in the United States.

APPENDIX A. EFFECTS ON THE ENVIRONMENT

Various observers are concerned that the development of shale resources may reduce the availability of water for other uses or contaminate it. They also have concerns, as well as hopes, about the effects of shale production on greenhouse gas emissions. The Congressional Budget Office (CBO) has not attempted to predict the future environmental consequences of shale development: The data about those consequences so far are not comprehensive enough, future consequences could differ from past ones because of the increasing scale of shale development, and future technological and regulatory developments are unclear. Instead, this appendix discusses the environmental effects of shale development on the basis of research and experience to date.[1] It also examines whether the federal government or state or local governments would be more likely to make economically efficient decisions about managing those effects.

Water Availability

Because hydraulic fracturing typically requires large quantities of water, shale development can contribute to strains on freshwater supplies. To date, conflicts between shale development and other uses of water have not been widespread, but some local ones have occurred.[2] In April and July of 2012, for example, after the flow volumes of local streams dropped below predetermined levels because of below-normal precipitation, the Susquehanna River Basin Commission temporarily suspended some arrangements for water withdrawals, most of which were related to shale development. Such conflicts may intensify as shale development increases. One study estimates that freshwater withdrawals by the oil and gas industry in Texas's Haynesville-Bossier Shale will reach more than 3 billion gallons annually between 2020 and 2035— which is more than double the 2010 amount and corresponds to about 6 percent of total current water use in the area.[3]

Shale developers can reduce the amount of freshwater that they require by reusing it. However, the potential for reusing water at a particular site depends on how much flows back from the well, the cost of treating it so that it can be reused, and the freshwater sources and disposal options available in the area. For example, in the Marcellus Shale, where underground disposal wells for used water are scarce, some operations reuse nearly all of the water that flows back from their wells—but doing so reduces their freshwater needs by 30 percent at most, because much of the water that is pumped into the ground remains there.[4] Replacing freshwater in hydraulic fracturing with treated, nonpotable water or hydrocarbon-based fluids could also reduce demands on freshwater resources, but it is too soon to know whether such technological developments will prove widely effective and cost-competitive.

In general, rights to use freshwater are controlled by states and are more limited than typical property rights, so water is not bought and sold in a free market.[5] For example, some state laws prevent those whose use returns water to the local environment from selling their water rights to those whose use does not do so. During water shortages, other state laws may subject all holders of water rights to proportional reductions in use; alternatively, more recent holders may be required to reduce their use so that those who preceded them in obtaining rights to the same source of water can claim their full allocation. Because of such limitations, the amount of water that shale operations use may be smaller than the amount that would maximize the benefit of such water use to society. On the other hand, shale operations may have access to more water than the most socially beneficial amount if their impact on the environment is not appropriately regulated or reflected in market prices (say, by charges that cover wastewater treatment costs and other environmental impacts).

Water Quality

Concerns about the effect of shale development on water quality involve various sources of potential contamination: sediment from the construction of drilling platforms; drilling fluids; various chemicals, which constitute up to 2 percent of the fracturing fluid injected into a well; the liquid removed from a well, which can include not only the fracturing fluid but also material from the shale formation (such as saltwater, organic compounds, heavy metals, and radioactive substances); and the extracted shale resources themselves.

Some of those contaminants might affect surface water, some might affect groundwater, and some might affect both (see Table A-1). However, certain routes of potential contamination—inadequate cleaning of the liquid removed from a fractured well before wastewater treatment facilities discharge it to surface water, for example, or underground migration from disposal wells to groundwater—are less likely than others, in part because they are regulated under the federal Clean Water Act of 1972 (CWA) or the federal Safe Drinking Water Act of 1974 (SDWA). In some cases, routes of potential contamination not currently regulated by the federal government could be addressed by regulations promulgated under state and local laws, especially those governing oil and gas production; in other cases, they already are, with varying degrees of stringency.[6]

Other provisions of the SDWA focus not on blocking contamination routes but on setting maximum concentrations of certain contaminants in water distributed by public drinking-water systems. Those provisions do not currently cover any of the 59 fracturing-fluid additives that companies have disclosed using routinely; however, limits for three of them—acetaldehyde, ethylene glycol, and methanol—are under consideration by the Environmental Protection Agency (EPA).[7] The SDWA concentration limits do not apply to private wells of drinking water, which serve about 15 percent of the U.S. population.

Of course, regulations may not be effective in meeting their stated goals, or they may meet their goals but at excessive cost. Thus, an important question to ask about the regulations related to shale development is whether they have positive net benefits—that is, benefits (which depend partly on the extent of compliance) minus costs (which include enforcement costs). A second important question is whether those net benefits could be increased by making the regulations more or less stringent. A third is whether the regulations reflect adequate scientific understanding of the risks associated with the larger-scale shale development anticipated for the future. At the request of the Congress, the EPA is writing a report on the potential effects of hydraulic fracturing on resources of drinking water; it should provide information relevant to those questions.[8]

Greenhouse Gas Emissions

Some observers are hopeful that shale development will result in lower overall greenhouse gas emissions because burning natural gas releases less carbon dioxide, a greenhouse gas, than burning other fossil fuels does. Specifically, when measured per unit of energy output, carbon dioxide emissions from natural gas combustion are about 45 percent lower than from coal combustion and about 30 percent lower than from oil combustion.[9] However, the effect on greenhouse gas emissions of replacing coal or oil with shale gas depends not simply on the fuels' different emissions during combustion but also on their different emissions during extraction, transport, processing, and distribution.[10] Moreover, shale gas that does not displace coal or oil almost certainly increases total emissions unless technology is used to control the emissions. And emissions from all fossil fuels may increase as shale development raises gross domestic product and increases the demand for energy.

Natural gas can more easily substitute for coal than for oil, because both gas and coal are commonly used to generate electricity. In fact, gas has been replacing coal in electricity generation for decades; that trend is expected to continue, partly because of the lower gas prices resulting from shale development.

In contrast, there is little current potential for natural gas to replace oil, whether as transportation fuel or for heating: Few vehicles run on natural gas, and only about 10 percent of heating is fueled by oil (some of which, moreover, occurs in areas where gas lines do not exist).[11]

The effects on greenhouse gas emissions of substituting shale gas for coal are difficult to estimate. A key factor is emissions of methane, the primary component of natural gas, during the initial phases of shale gas extraction. Methane is a more powerful greenhouse gas than carbon dioxide; the latest report of the Intergovernmental Panel on Climate Change estimates that methane's impact on the climate, per unit of mass, is 84 times greater than carbon dioxide's over a period of 20 years and 28 times greater over 100 years.[12] The quantity of methane emissions depends critically on how producers handle the gas that emerges as the fracturing fluid injected into a well returns to the surface before the main extraction of shale energy. They might release it into the atmosphere; burn it off to reduce its climate impact, a process called flaring; or capture it with "green completion" technologies, which reduce total emissions by 90 to 95 percent and are the most effective way to minimize their impact on the climate.[13]

The relative use of those three options is not well documented, and estimates vary widely.[14] In October 2012, the federal government began requiring shale gas developers either to use green completions or to flare their emissions.[15] By January 2015, green completions will be required for new hydraulic fracturing at gas wells, although a few categories of wells, such as those used to look for gas rather than to extract it, will still be allowed to flare.[16] There are no such requirements for oil wells.

Table A-1. Possible Routes of Water Contamination by Shale Development

Contamination Route	Applicable Federal or State Regulation	Additional Information
Surface Water		
Sediment transported by storm water	Such contamination is regulated by CWA, but only if it violates a water quality standard; it is also regulated by some states.[a]	
Spills, overflows, and seepage from storage pits and tanks	CWA requires oil and gas producers to have plans to prevent and contain certain spills; in some cases, it requires discharge permits and pollution prevention plans.[b]	Some spills have occurred because of equipment failures that developers, viewing them as unlikely, had not addressed in their prevention plans.
Inadequate cleaning of flowback water—which consists of fracturing fluid and fluid from the rock formation that surface after hydraulic fracturing—by wastewater treatment facilities	Commercial and industrial sources of wastewater that would pose problems for wastewater treatment facilities are required by general standards in CWA regulations to pretreat their wastewater. Specific CWA standards for the pretreatment of flowback water from shale gas wells are expected to be proposed.[c] Some states also regulate the discharge of flowback water to wastewater treatment facilities.	Shale operations may not fully comply with pretreatment requirements. For example, CWA violations in the Allegheny River watershed between 2007 and 2011 were associated with wastewater from the Marcellus Shale processed at three Pennsylvania wastewater treatment plants between 2007 and 2011. Since mid-2011, because of a combination of state prohibitions and voluntary actions, shale operators have generally not sent flowback to Pennsylvania wastewater facilities that cannot provide pretreatment.

Contamination Route	Applicable Federal or State Regulation	Additional Information
		Outside the Marcellus Shale area, underground disposal wells are more widely available, and developers therefore have less economic incentive to dispose of flowback water through wastewater treatment plants.
Groundwater		
Spills	CWA regulations designed to protect surface water may also protect groundwater.[d]	
Underground migration from rock formations targeted by developers	SDWA regulations apply to hydraulic fracturing only in the cases (about 2 percent of the total) in which the fracturing fluid includes diesel.[e]	Typically, the target formation is separated from sources of groundwater used for consumption by thousands of feet of rock.[f] Some exceptions may exist: A preliminary Environmental Protection Agency report found that some hydraulic fracturing operations in Wyoming that occurred less than 500 feet below depths reached by drinking-water wells may have contaminated deeper portions of the groundwater aquifer that the wells drew from.[g]

Table A-1. (Continued)

Groundwater (Continued)		
Underground migration from leaking wells	The federal government does not regulate well integrity; some states do.	Methane contamination of some drinking-water wells in Pennsylvania and Texas has been linked to leakage from hydraulically fractured shale gas wells.h Also, in 2011, Pennsylvania's Department of Environmental Protection fined a shale gas developer for contaminating well water. Water may contain methane for reasons unrelated to hydraulic fracturing, so it can be difficult to assign responsibility unless water samples were taken before shale development began.
Underground migration from disposal wells	Disposal wells require an SDWA permit.[i]	SDWA permits are issued after the government has determined that the rock formation where a disposal well will be located is sufficiently isolated from groundwater.

Source: Congressional Budget Office.

Note: CWA = Clean Water Act; SDWA = Safe Drinking Water Act.

[a] Sec. 303 of the Clean Water Act of 1972, Public Law 92-500 (codified at 33 U.S.C. §1313 (2012)).

[b] Sections 301, 311, 402, and 404 of the Clean Water Act, P.L. 92-500 (codified at 33 U.S.C. §§1311, 1321, 1342, and 1344 (2012)); 40 C.F.R. §117 (2013).

[c] Clean Water Act; 40 C.F.R. §437 (2013); and Notice of Final 2010 Effluent Guidelines Program Plan, 76 Fed. Reg. 66302 (October 26, 2011).

[d] Sec. 340 of the Clean Water Act, P.L. 92-500 (codified at 33 U.S.C. §1314 (2012)).

[e] Sec. 1421 of the Safe Drinking Water Act of 1974, P.L. 93-523 (codified at 42 U.S.C. §300h(d) (2012)); and Mike Soraghan, "Hydraulic Fracturing: Diesel Still Used to 'Frack' Wells, FracFocus Data Show," *EnergyWire* (August 17, 2012), http://tinyurl.com/puduv5m.

[f] See George E. King, "Hydraulic Fracturing 101" (paper presented at the Society of Petroleum Engineers Hydraulic Fracturing Technology Conference, The Woodlands, Texas, February 6–8, 2012), http://tinyurl.com/nt3r3w7 (PDF, 7 MB); Stephen G. Osborn and others, "Methane Contamination of Drinking Water Accompanying Gas-Well Drilling and Hydraulic Fracturing," *Proceedings of the National Academy of Sciences*, vol. 108, no. 20 (May 17, 2011), pp. 8172–8176, http://tinyurl.com/5w227nj; and Nathaniel R. Warner and others, "Geochemical Evidence for Possible Natural Migration of Marcellus Formation Brine to Shallow Aquifers in Pennsylvania," *Proceedings of the National Academy of Sciences*, vol. 109, no. 30 (July 24, 2012), pp. 11961–11966, http://tinyurl.com/ckfheor.

[g] Dominic C. DiGiulio and others, *Investigation of Ground Water Contamination Near Pavillion, Wyoming*, EPA 600/R-00-000 (draft, Environmental Protection Agency, December 2011), http://go.usa.gov/KGNG (PDF, 15 MB). "Less than 500 feet" is a CBO conversion from metric data on page xi of the report.

[h] Thomas H. Darrah and others, "Noble Gases Identify the Mechanisms of Fugitive Gas Contamination in Drinking-Water Wells Overlying the Marcellus and Barnett Shales," *Proceedings of the National Academy of Sciences*, vol. 111, no. 39 (September 13, 2014), pp. 14076– 14081, www.pnas.org/content/111/39/14076.

[i] Safe Drinking Water Act; 40 C.F.R. §144.31 (2013).

Because of differences in production methods, shale gas that substitutes not for coal or oil but for gas from other sources increases total emissions of greenhouse gases when emission controls are not in place during the drilling and extraction phases.[17] Subsequent production activities, such as transport, processing, and distribution, may be even more significant sources of methane emissions, but they are essentially the same for shale gas and conventional gas.[18]

A given volume of shale gas increases greenhouse gas emissions even more when it substitutes for energy sources other than fossil fuels—such as nuclear plants, windmills, and solar panels—because those energy sources emit no greenhouse gases at all in use. (A comprehensive comparison of those energy sources with shale gas would include the emissions associated with the construction of facilities, energy production, processing, and transport.) Similarly, when shale gas does not displace other energy sources but simply increases total energy use, all of the emissions resulting from its production, distribution, and use are net additions. All things being equal, such an increase in energy use would be the likely result of lower prices for natural gas and other forms of energy. For instance, families might choose to keep their homes warmer in the winter because the cost was lower; for the same reason, firms might reduce their investments in energy-efficient technologies. Also, the faster economic growth spurred by cheaper energy would increase demand for energy in general, including fossil fuels.

Environmental Policy in a Federal System of Government

There are a number of ways in which the government may influence the environmental effects of shale development: choosing standards for water use, water quality, and greenhouse gas emissions; deciding on the acceptable methods of meeting those standards; and funding related research. Federal policymakers may wish to consider whether the current division of regulatory responsibilities among federal, state, and local governments is likely to lead to decisions that maximize the net benefits to society.[19]

Standards

There is a stronger rationale for states and localities to set environmental standards for shale development if the costs and benefits of controlling the environmental effects of such development occur solely within their borders. The effects of shale development on the availability of water for other uses may be limited to a local area (which would not necessarily be a single local jurisdiction), or they may extend to a broader region—for instance, by affecting groundwater levels in regional aquifers. Similarly, the effects of shale development on water quality may be confined to a local area or extend beyond state boundaries. Rising or falling greenhouse gas emissions have global effects.

Other considerations include which level of government has the most information about underlying costs and benefits; whether centralizing the process of setting standards would yield savings in administrative costs; and the objectives and capabilities of different levels of government. For example, federal policymakers might choose standards that gave greater weight to environmental costs than state standards would, because states' objectives include competing with each other for industries and jobs.

Methods of Meeting Those Standards

There is a stronger rationale for a state or local role, rather than a federal one, in deciding which methods may be used to meet environmental standards if the opportunities and costs of available methods vary among areas. For example, the cost of addressing water quality concerns associated with hydraulic fracturing can vary by locality, depending in part on whether local geology allows producers to dispose of wastewater in underground disposal wells; a federal decision to require that method of disposal might therefore be overly costly in some areas.

Another consideration is whether a particular method of meeting environmental standards would be more cost-effective if it was put to use on a large scale; if so, the argument for federal regulation is stronger. That argument is also stronger when a method of meeting environmental standards would have effects outside the state in which it was used. And constraints on states' willingness and ability to select efficient methods of meeting standards would likewise argue for federal regulation, just as such constraints on local governments would argue for state regulation.

Research

If many states face the same type of environmental problem, a stronger rationale exists for the federal government to determine and fund a research agenda related to that problem. An example is research to determine precisely how much methane is emitted by the development of shale gas, because such research would help inform policies on greenhouse gas emissions in many states and at the federal level. If the endeavor was left to the states, some studies that would be worthwhile to the nation as a whole might not be undertaken, because they would not be justified by the benefits to a single state or even a small group of them. Further, research conducted by one state might be duplicated by another if the states failed to coordinate plans or share findings.

Appendix B. The Basis of CBO's Estimates of Longer-Term Effects on Economic Output

The Congressional Budget Office (CBO) estimates that real (inflation-adjusted) gross domestic product (GDP) will be 0.7 percent higher in 2020 and 0.9 percent higher in 2040 than it would have been without the development of shale resources. The analysis underlying those estimates involved two main steps:

- CBO compared a recent energy market projection by the Energy Information Administration (EIA) in which shale resources were available in the United States to an alternative projection by CBO in which those resources were not available. CBO estimated energy prices and quantities for that alternative projection by extrapolating from EIA's estimates of the effects on energy markets of differences in future amounts of shale energy production.
- Using the two sets of projections, CBO estimated the impact that different quantities and prices in energy markets would have on GDP, focusing on increases in the productivity of existing labor and capital and increases in the amount of labor and capital in use.

Oil and Natural Gas Markets with and without Shale Resources

CBO analyzed the domestic consumption and net exports of oil products (defined here to include fuels derived from petroleum as well as other liquid fuels, such as ethanol, biodiesel, and natural gas plant liquids) and of natural gas in the projections with and without shale resources, as well as the market prices of oil and natural gas in those two projections. The analysis also took account of the profitability of shale development in the projection that included shale resources. In addition, CBO constructed alternative cases to account for uncertainty about future production levels, energy prices, and profitability of shale gas production.

Consumption and Net Exports of Oil and Natural Gas with and without Shale Resources

The absence of shale resources would reduce the overall domestic production of oil and natural gas, as well as the domestic use and net exports of natural gas and oil products. However, the magnitudes of those effects differ: CBO estimates that 60 percent of a reduction in the amount of domestic shale gas produced would be reflected in lower domestic gas consumption, that roughly 5 percent would be met by an increase in the production of natural gas not from shale, and that the remaining 35 percent would be reflected in lower exports or higher imports. In contrast, CBO estimates that only 10 percent of a reduction in the amount of tight oil produced would take the form of lower domestic oil consumption, that roughly 20 percent would reflect an increase in the domestic production of crude oil from other sources, and that about 70 percent would be absorbed by greater net imports.[1]

Those values are averages of the values resulting from EIA's estimates of the sensitivities of future consumption levels to the availability of shale energy production.[2] Differences in market size explain the different effects. Because overseas transport to and from North America is more costly for natural gas than for crude oil, U.S. production of shale gas has a comparatively large effect on North American gas prices and thus a comparatively large effect on U.S. gas consumption. In contrast, U.S. production of tight oil has a comparatively small effect on world oil prices, so changes in that production have a relatively small effect on domestic oil consumption and are primarily reflected in changes in net exports.[3]

Thus, given EIA's baseline projection that shale gas production will be about 9.5 trillion cubic feet (Tcf) in 2014, CBO projects that in the absence of that production, total domestic gas production would be about 9.0 Tcf lower (because conventional production would increase by about 0.5 Tcf), domestic consumption would be about 5.5 Tcf lower (roughly 60 percent of 9.5 Tcf), and net exports would be about 3.5 Tcf lower (through lower exports or higher imports). CBO similarly estimates that the domestic production, consumption, and net exports of oil products would all be lower in 2014— by 3.9 million, 0.5 million, and 3.4 million barrels per day, respectively—in the absence of the estimated 4.8 million barrels per day of liquid fuels attributable to shale development.[4] CBO's projections of the production and consumption of oil and natural gas with and without shale resources are shown in Figure B-1.

Market Prices of Oil and Natural Gas with and without Shale Resources

Given the estimated differences in the domestic consumption of oil products and natural gas with and without shale resources, the differences in market prices supporting those consumption levels can be calculated by using the elasticity of demand for those fuels. Elasticities measure the percentage change in the production or the consumption of a good for each 1 percent change in the price. Given the estimated reduction in consumption when shale resources are not available (measured relative to the baseline projection, in which they are available), the percentage difference in market prices ($\% \Delta P$) is given by the following equation:

$$\%\Delta P = \frac{\%\Delta D}{\varepsilon_d}$$

In that equation, $\% \Delta D$ refers to the percentage difference in the consumption of oil products or natural gas, and ε_d is the elasticity of demand with respect to differences in the market price of oil or gas.

On the basis of EIA's most recent long-term outlook, CBO estimates the long-term elasticities of demand for oil products and for natural gas to be about -0.5 each, so that a 20 percent increase in the price of oil products or natural gas would reduce the amount of oil and gas consumed by 0.5 times as much, or 10 percent. For instance, domestic gas consumption in 2040 would be about 11.5 Tcf lower in the absence of shale supplies, CBO estimates; that 11.5 Tcf is about 35 percent of 2040 domestic consumption in the baseline projection, implying that gas prices (given the elasticity of demand) would be about 70 percent higher without shale supplies, as Figure B-2 shows. Again, U.S. production of tight oil will have a smaller effect on world prices, which would be about 5 percent higher otherwise, because U.S. tight oil as a share of world liquid fuel supplies is much smaller than U.S. shale gas as a share of North American gas supplies.

Combining those estimated price effects with estimates of production volumes, CBO calculates that the value of U.S. oil and gas production in 2020 in the absence of shale development would be about $495 billion (measured in 2012 dollars), as opposed to $645 billion with shale development.[5] In 2040, the value of U.S. oil and gas production would be roughly $760 billion without shale development, as opposed to $950 billion with it.

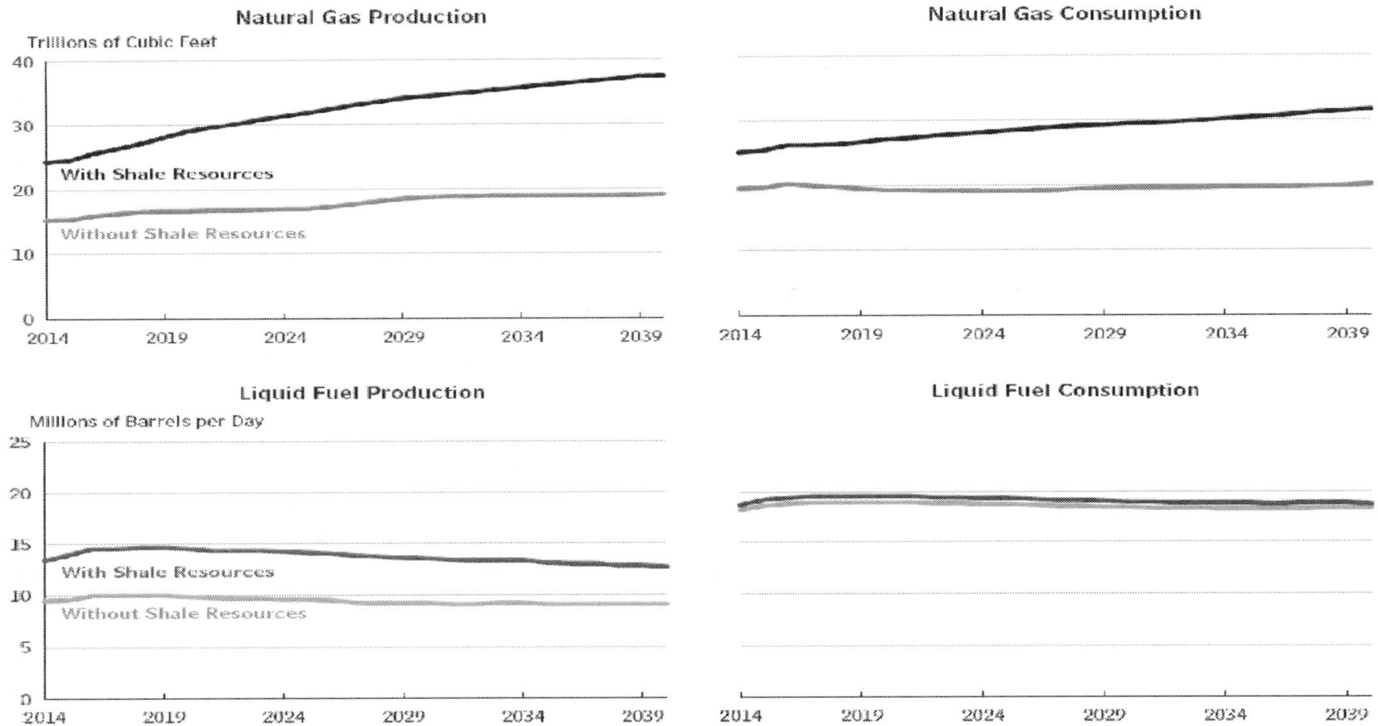

Source: Congressional Budget Office based on Energy Information Administration, *Annual Energy Outlook 2014 With Projections to 2040*, DOE/EIA-0383(2014) (April 2014), http://go.usa.gov/8KyF (PDF, 12 MB).

Notes: Production and consumption amounts for natural gas and liquid fuels when shale resources are present (labeled “With Shale Resources”) are the Energy Information Administration’s most recent long-term projections. Projections when shale resources are not present (labeled “Without Shale Resources”) are CBO’s estimates.

The category “Liquid Fuel” includes crude oil, biofuels, natural gas plant liquids, and other liquid fuels.

The projections for all years are based on the assumptions that the economy is producing close to its maximum sustainable level of output and that energy markets are stable. As the text explains, CBO expects that the actual effects would be somewhat different in the short term.

Figure B-1. Effects of Shale Resources on the Domestic Production and Consumption of Natural Gas and Liquid Fuels.

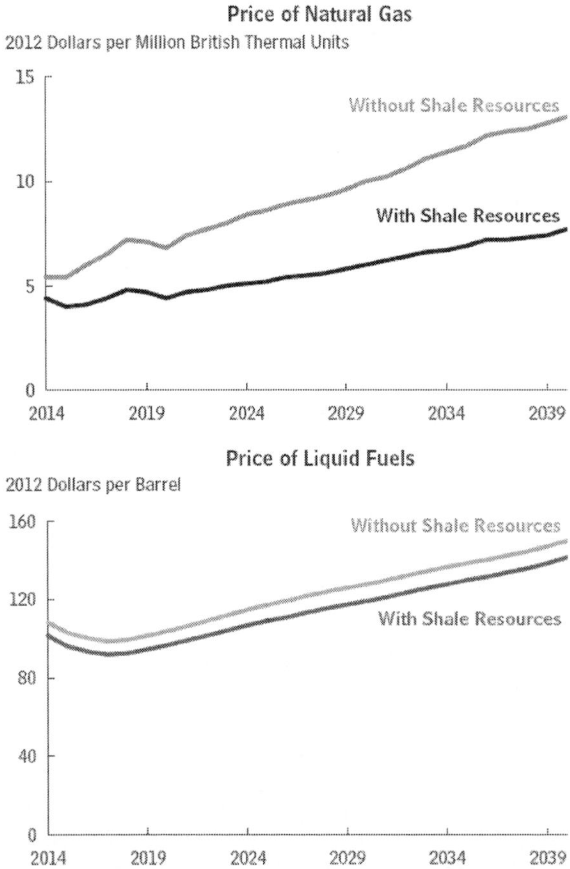

Source: Congressional Budget Office based on Energy Information Administration, Annual Energy Outlook 2014 With Projections to 2040, DOE/EIA-0383(2014) (April 2014), http://go.usa.gov/8KyF (PDF, 12 MB).

Notes: Market prices for natural gas and liquid fuels when shale resources are present (labeled "With Shale Resources") are the Energy Information Administration's (EIA's) most recent long-term projections. Projections when shale resources are not present (labeled "Without Shale Resources") are CBO's estimates. CBO reports prices in 2012 dollars because that was the basis that EIA used when modeling its projections in real (inflation-adjusted) terms.

The category "Liquid Fuels" includes crude oil, biofuels, natural gas plant liquids, and other liquid fuels.

The projections for all years are based on the assumptions that the economy is producing close to its maximum sustainable level of output and that energy markets are stable. As the text explains, CBO expects that the actual effects would be somewhat different in the short term.

Figure B-2. Effects of Shale Resources on the Price of Natural Gas and Liquid Fuels.

Profitability of Shale Development

Excess returns from producing shale resources—that is, revenues less production costs—contribute to GDP. They represent the difference between the output of labor and capital when used to produce shale resources and the output of that labor and capital when used elsewhere in the economy.[6] Those excess returns are determined by the volume of tight oil and shale gas expected to be produced in future years and the difference between the market prices of oil and natural gas and the average break-even cost of producing shale resources—that is, the lowest average price necessary for developers of shale resources to cover their costs of labor and capital.

CBO's estimates of the current and future break-even costs of tight oil and shale gas production are based on recent estimates from the International Energy Agency (IEA).[7] IEA estimates that current production costs of tight oil worldwide range from $60 to $100 per barrel (in 2013 dollars). CBO estimates that the average break-even cost of U.S. tight oil is the midpoint of IEA's range of world costs—that is, $80 per barrel. That figure is consistent with the fact that the production of U.S. tight oil became significant in the last few years, as world oil prices climbed past $100 per barrel. For shale gas, CBO estimates that the current break-even production cost in the United States is $3 per million British thermal units (mmBtu), at the low end of IEA's worldwide range of $3 to $10 per mmBtu (in 2013 dollars). One reason for that estimate is that investment in shale gas in the United States has remained robust even as gas prices have fallen below $4 per mmBtu. Another reason is that world production of shale gas has been concentrated in the United States; domestic production costs are probably lower than in countries where development is proceeding more slowly.

CBO expects that break-even costs will grow in real terms as real prices for oil and gas grow. As market prices rise, companies will develop shale gas and tight oil that are more costly to produce, thereby raising the average cost of production. Specifically, CBO models the average break-even cost for tight oil as growing at the same rate at which real prices for crude oil do, so that the inflation-adjusted break-even cost grows from $80 per barrel in 2014 to about $110 per barrel by 2040. For shale gas, however, CBO models the inflation-adjusted break-even cost as growing at half the rate expected for the real price of natural gas, so that the real break-even cost grows from $3 per mmBtu in 2014 to about $4 per mmBtu by 2040. CBO's different expectations for shale gas and tight oil are broadly consistent with EIA projections that shale gas production will grow as a share of total U.S. gas production in coming decades because it will become relatively cheaper, while tight oil's share of total U.S. production of liquid fuels will not change significantly.

CBO estimates that inflation-adjusted excess returns in 2020 will total roughly $20 billion for shale gas, an estimate based on about 13 Tcf of production, a price of $4.40 per mmBtu, and a break-even cost of $2.90 per mmBtu. CBO also estimates—on the basis of about 5.6 million barrels per day of expected production, a price of $97 per barrel, and a break-even cost of $75 per barrel—that inflation-adjusted excess returns in 2020 will total $45 billion for tight oil. In 2040, CBO expects inflation-adjusted excess returns to total about $75 billion for shale gas and $50 billion for tight oil.

Uncertainty About Projections

Shale energy production, market prices of gas and oil, and the profitability of shale gas production may be significantly higher or lower than CBO projects in its baseline. All things being equal, higher production, market prices, and profitability would mean larger effects on GDP. To illustrate the uncertainty accompanying its baseline estimates, CBO generated alternative projections of those three factors.

CBO constructed a range of shale production quantities on the basis of a recent EIA "low-resource" projection in which the total amount of gas and oil recoverable from each shale gas and tight oil well was 50 percent lower than in EIA's baseline projection.[8] Comparing the two projections, CBO calculated the percentage difference in the number of Btus of shale gas and tight oil produced each year. CBO then obtained its range of shale production quantities by increasing or decreasing its baseline projections for shale gas and tight oil production by those year-by-year percentages.[9] The resulting percentage deviations from baseline production levels are shown in Figure B-3.

To generate alternative projections of market prices, CBO relied on EIA's most recent long-term outlook, which includes projections for higher and lower oil prices. (CBO used the market price of crude oil to approximate the price of liquid fuels, which include crude oil, the petroleum products produced from it, and other liquids, such as biofuels and natural gas plant liquids.) In the absence of analogous EIA projections for natural gas prices, CBO used the same percentage increase and decrease (relative to the baseline projections) that it did for the price of oil. The high and low prices that CBO obtained for natural gas and oil are shown in Figure B-4.

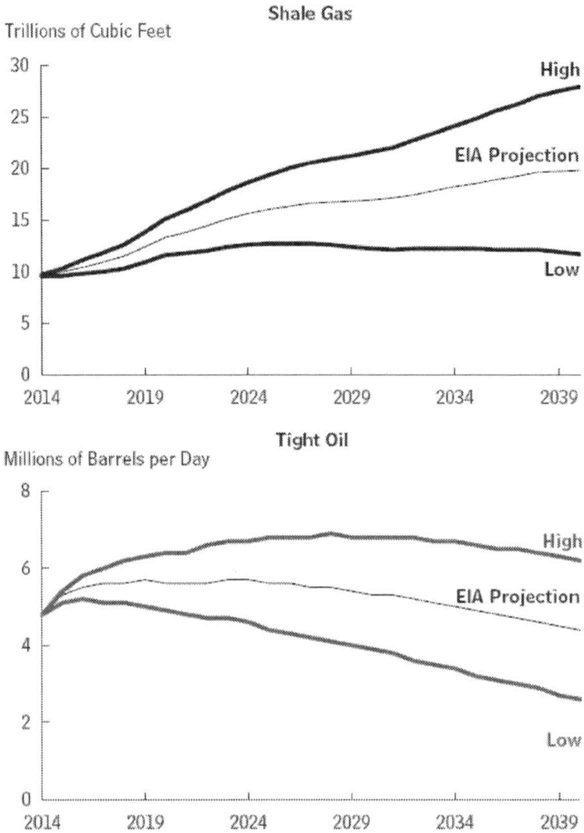

Source: Congressional Budget Office based on Energy Information Administration, Annual Energy Outlook 2014 With Projections to 2040, DOE/EIA-0383(2014) (April 2014), http://go.usa.gov/8KyF (PDF, 12 MB).

Notes: Here, the production of tight oil includes not only crude oil that is extracted from shale by means of hydraulic fracturing but also CBO's estimate of the portion of the production of natural gas plant liquids—forms of natural gas that substitute for certain petroleum products—that is produced by hydraulic fracturing.

EIA = Energy Information Administration.

Figure B-3. Projected High and Low Production of Shale Gas and Tight Oil.

To measure the sensitivity of CBO's findings to the profitability of producing shale gas, CBO considered cases in which break-even costs for shale gas grew at 25 percent and 75 percent of the yearly change in gas prices, rather than the 50 percent assumed in the baseline projection.

In dollar terms, excess returns from shale gas production are greatest if the market price of gas is high and the break-even cost of production is low; conversely, excess returns are smallest if the market price of gas is low and the break-even cost is high.

When evaluating the effects of greater or lesser availability of shale energy supplies, CBO included price effects resulting from those differences in supplies. For any given assumption about other factors affecting prices—whether those factors lead to high prices, baseline prices, or low prices—more abundant shale energy supplies will, all else being equal, reduce those prices. Similarly, those prices will be higher if supplies are less abundant. CBO used the same approach that was outlined above to calculate the effect of more or less shale production on those price scenarios: More (or less) abundant shale gas or tight oil boosts (or lowers) consumption levels, leading to a percentage change in market prices that is calculated by means of the elasticity of demand.

Effects of Shale Development on Economic Output in the Longer Term

Real GDP will be higher in the longer term than it would have been without the development of shale resources. CBO estimates that, by increasing the productivity of labor and capital, the production and use of shale gas will make GDP about 0.2 percent higher in 2020 and about 0.4 percent higher in 2040 than it would have been otherwise, and the production and use of tight oil will make GDP about 0.2 percent higher in both of those years (see Table B-1). Moreover, because that higher productivity of labor and capital will induce a greater supply of labor and capital in the economy, shale development will further increase GDP by roughly 0.3 percent and 0.4 percent in 2020 and 2040, respectively. All told, CBO's baseline long-term projection for real GDP is 0.7 percent higher in 2020 and 0.9 percent higher in 2040 than it would have been without the development of shale resources.[10]

CBO's analysis focused on the effects on GDP in the longer term—that is, after the economy moves back toward producing its maximum sustainable level of output. In the near term, the increase in GDP associated with increased production and use of shale resources is greater because the firms producing and using more shale resources use some labor and capital that would otherwise have been underused. For example, in the current economic environment, some of the workers employed by businesses engaged in hydraulic fracturing would otherwise have been unemployed. In the economic environment that CBO expects in the long run, however, such workers would otherwise have been employed in other jobs.

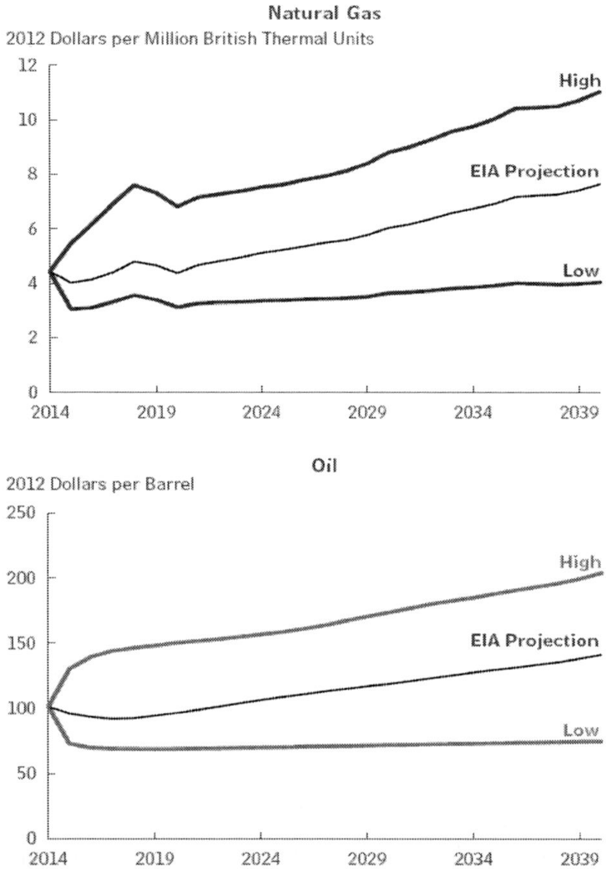

Source: Congressional Budget Office based on Energy Information Administration, *Annual Energy Outlook 2014 With Projections to 2040*, DOE/EIA-0383(2014) (April 2014), http://go.usa.gov/8KyF (PDF, 12 MB).

Notes: Here, the production of tight oil includes not only crude oil that is extracted from shale by means of hydraulic fracturing but also CBO's estimate of the portion of the production of natural gas plant liquids—forms of natural gas that substitute for certain petroleum products—that is produced by hydraulic fracturing.

EIA = Energy Information Administration.

Figure B-4. Projected High and Low Market Prices of Natural Gas and Oil.

Effects of Shale Gas on the Productivity of Labor and Capital

To think about the long-term effects of shale gas on productivity and hence on GDP, consider Figure B-5 on page 39, which shows two hypothetical supply curves for natural gas: one that does not include shale gas and one that does. The supply curve without shale gas is line S1, and it intersects the demand curve for natural gas—line D—at point A, showing that without shale gas, the market for natural gas would clear (that is, demand would equal supply) at a price of $4 per thousand cubic feet (mcf).[11] Once shale gas becomes available, the total supply of natural gas shifts to S2, meaning that more gas is available at a lower price. The horizontal difference between S1 and S2 is the amount of shale gas supplied at a given price. For example, at a price of $3 per mcf, the supply of shale gas is the horizontal difference between the quantities represented by points E and B. The market now clears at point B, at a price of $3.

The gain in GDP in the long run from that outward shift of the supply curve closely corresponds to the area enclosed by points A, B, and C. The gain is composed of three parts:

- The gain from the increased productivity of labor and capital used to produce shale gas, which corresponds to the area enclosed by points B, C, and E;
- The gain in productivity from producing shale gas instead of more expensive conventional gas, which corresponds to the triangle with corners at points A, E, and F; and
- The gain in productivity from the increased consumption of gas by domestic businesses and households, which is included in the area enclosed by points A, B, and F.[12]

Table B-1. Effects of U.S. Shale Development on GDP

Percent	**Effect on GDP**	
	2020	**2040**
Increased Productivity of Labor and Capital		
Shale gas		
Gain in productivity of labor and capital producing shale gas	0.1	0.3
Gain in productivity from producing shale gas instead of conventional gas	*	*

Table B-1. (Continued)

Percent	**Effect on GDP**	
	2020	**2040**
Gain in productivity from increased consumption of gas	0.1	0.1
Subtotal	0.2	0.4
Tight oil		
Gain in productivity of labor and capital producing tight oil	0.2	0.2
Gain in productivity from producing tight oil instead of conventional oil	*	*
Gain in productivity from increased consumption of oil	*	*
Subtotal	0.2	0.2
Total	0.4	0.5
Additional Supply of Labor and Capital	0.3	0.4
Total Effect of U.S. Shale Development on GDP	**0.7**	**0.9**
Memorandum:[a]		
Gains in Productivity of Labor and Capital Producing Shale Resources	0.3	0.4
Gains in Productivity From Producing Shale Resources Instead of Conventional Resources	*	*
Gains in Productivity From Increased Consumption of Gas and Oil	0.1	0.1

Source: Congressional Budget Office.

Notes: Totals may not match because of rounding.

Tight oil is crude oil extracted from shale and certain other dense rock formations by means of hydraulic fracturing. GDP = gross domestic product; * = between zero and 0.05 percent.

a. These lines add shale gas's effects on GDP (shown above in the table) to tight oil's effects on GDP (also shown above in the table).

Gains from the Increased Productivity of Labor and Capital Producing Shale Gas

In the longer run, the development of shale gas leads to higher GDP in part because labor and capital can be used more productively to produce shale gas than to produce other output. In Figure B-5, that gain is shown by the area between the two supply curves and below $3 per mcf—the triangle bounded by points B, C, and E. In Figure B-6, the same gain is shown by the area above the supply curve S3— which is a supply curve for shale gas alone—and below $3; it is also a triangle bounded by points B, C, and E. (The distance from B to E in Figure B-5 represents the quantity of shale gas produced at the new equilibrium price—the same quantity represented by the distance from B to E in Figure B-6. The areas of the two triangles bounded by points B, C, and E in the two figures are the same because their bases and heights are the same.)

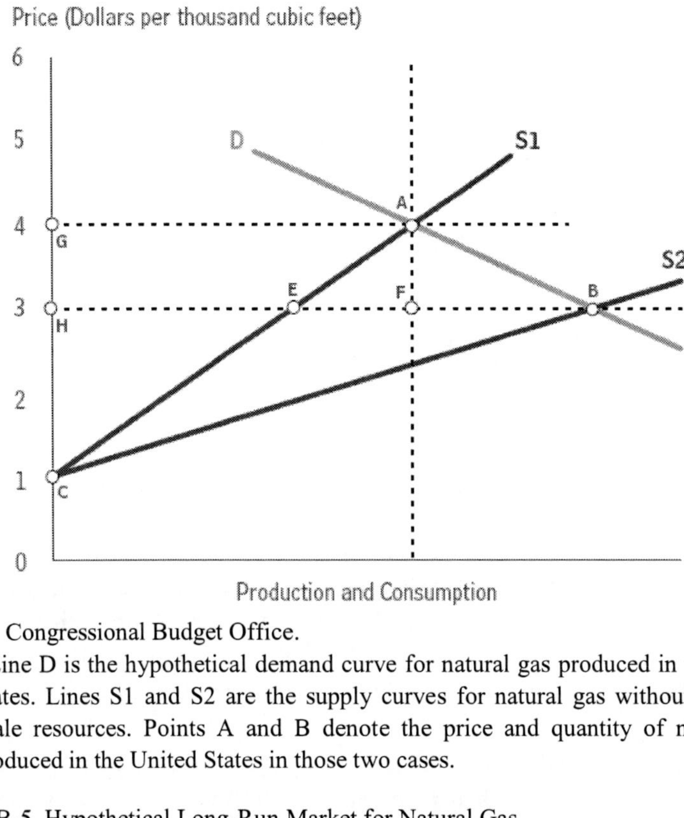

Source: Congressional Budget Office.

Note: Line D is the hypothetical demand curve for natural gas produced in the United States. Lines S1 and S2 are the supply curves for natural gas without and with shale resources. Points A and B denote the price and quantity of natural gas produced in the United States in those two cases.

Figure B-5. Hypothetical Long-Run Market for Natural Gas.

At any point along S3 in Figure B-6, the addition to GDP from the production of shale gas is the vertical distance between the supply curve and the price of gas. For example, at point C, a firm is willing to supply shale gas at a long-run price of $1 per mcf or more because the cost of labor and capital used by that firm to produce 1 mcf of gas is $1.

That $1 of labor and capital would be producing $1 of GDP if employed in other industries. At a price of $3, the shift of labor and capital from other industries generates an extra $2 of GDP, the vertical difference between points C and E.

To take another example: At point B of Figure B-6, the production of an additional 1 mcf of shale gas is profitable only if the long-run price is $3 or more. Because the labor and capital used to produce that gas could produce $3 of GDP elsewhere, there is no net gain in GDP from the production of shale gas that is also valued at $3.

Another way to think about the gain in GDP from the production of shale gas is to subtract the total cost of producing shale gas from the total value of that gas. The cost is the quantity of gas produced multiplied by the break-even cost, and it is shown by the area enclosed by points B, C, J, and K in Figure B-6. The value of the gas produced is its price multiplied by the quantity produced, which is shown by the rectangle whose corners are points B, E, J, and K. The difference is represented by the triangle enclosed by points B, C, and E.

CBO estimated the long-term gain in GDP from the production of shale gas by multiplying its estimate of the amount of shale gas produced (which would correspond to the distance between points B and E in Figure B-6, though that figure, again, is hypothetical) by the difference between the price of that gas and CBO's estimate of the average break-even cost of that production (which would be equivalent to the vertical midpoint of the supply curve between points B and C).

CBO estimates the GDP gain from the production of shale gas to be 0.1 percent of GDP in 2020 and 0.3 percent in 2040. In 2012 dollars, the 2040 estimate is about $75 billion, which is based on projections that shale gas production in 2040 will total about 20 billion mcf; that the market price of gas will be about $8 per mcf; and that the average break-even cost of shale gas will be about $4 per mcf.

The gains in GDP from producing shale gas are greater in the near term than in the long term. When labor and capital are underused, some of the resources used to produce shale gas would not otherwise be producing GDP. In that case, the gain to GDP from producing shale gas is represented not only by the triangle enclosed by points B, C, and E in Figure B-6 but also by part of the area enclosed by points B, C, J, and K.

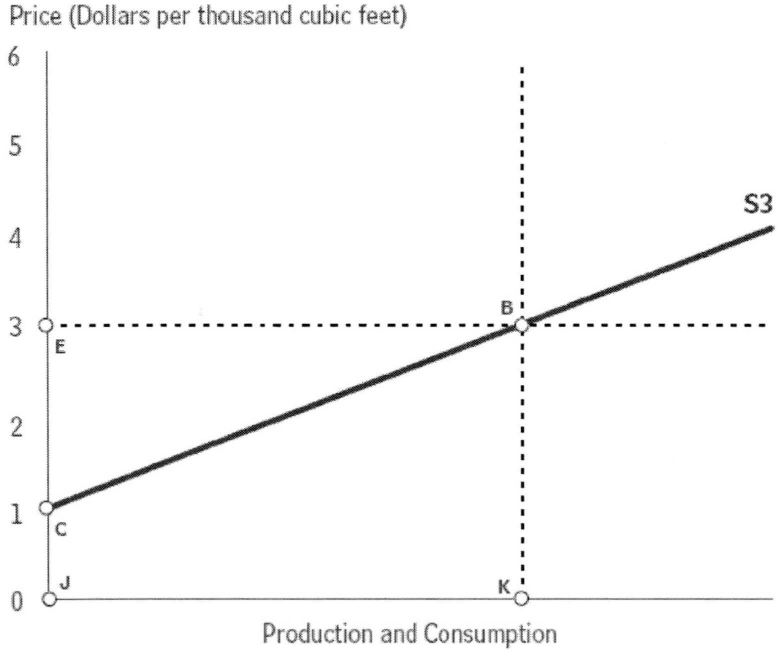

Source: Congressional Budget Office.

Note: Line S3 is the supply curve for shale gas. Point B denotes the quantity of shale gas produced in the United States at a price of $3 per thousand cubic feet.

Figure B-6. Hypothetical Long-Run Supply Curve for Shale Gas.

Gains in Productivity from Producing Shale Gas Instead of Conventional Gas

The gain in GDP from substituting shale gas for conventional gas that is no longer economical to produce because of the lower price of gas corresponds to the triangle enclosed by points A, E, and F in Figure B-5. A firm willing to supply additional conventional gas at point A requires a price of $4 per mcf because it uses $4 of labor and capital. Displacing that conventional gas with shale gas produced at a cost of $3 thus frees up $1 of labor and capital for other uses, increasing GDP by $1. A firm willing to supply additional conventional gas at point E uses $3 of labor and capital to produce that gas, so replacing it with shale gas at $3 per mcf does not add to GDP.

The savings that consumers of natural gas realize because of the fall in price from $4 to $3 is represented by the rectangle enclosed by points A, F, H, and G in Figure B-5. Most of those savings do not add to GDP but instead represent a transfer from producers to consumers of natural gas. Within that rectangle, only the savings in production costs, which are represented by the triangle outlined by points A, E, and F, add to GDP.

CBO estimated the gain in GDP from substituting shale gas for conventional gas by multiplying the estimated difference in U.S. production of conventional gas by one-half the difference between the projected price of gas and the estimated price that would prevail in the absence of shale resources. The gain in GDP is projected to be very small in both 2020 and 2040—less than 0.05 percent of GDP—because the production of conventional gas is expected to be only about 5 percent lower than it would have been without shale resources.

Gains in Productivity from Increased Consumption of Gas

The development of shale resources also raises GDP as consumption of cheaper gas frees up labor and capital for other uses, allowing the economy to produce a greater value of goods and services with the same total amount of labor and capital. In Figure B-5, those gains are included in the triangle enclosed by points A, B, and F. (That triangle also includes gains to foreign firms that use more natural gas. Those gains do not contribute to U.S. GDP, and CBO excluded them from its calculations.)

As the price of natural gas falls, productivity increases for two reasons. First, some firms are able to reduce their cost of producing goods and services by substituting cheaper gas for labor, capital, or other inputs. For example, an electric utility might generate more electricity from gas and less from coal. Second, the composition of output produced in the economy changes as households and firms shift toward goods and services that are gas-intensive and thus become relatively less costly to produce. For example, households might buy more tires, fertilizer, and plastic containers, and spend less on clothes.

In both cases, the benefit to the economy of each additional thousand cubic feet of natural gas used is the difference between the highest price at which that gas would be purchased, represented by the heights of the points along demand curve D in Figure B-5, and its actual selling price. Consider, for example, an electric utility willing to buy an additional 1 mcf of natural gas at a price up to $4—the level of demand represented by point A in the figure. That willingness to spend up to $4 reflects the utility's ability to substitute the gas—as well as the costs of using it—for other resources that together cost the same amount. For instance, the utility might buy 1 mcf of gas costing $4, plus $5 of other necessary goods and services, to generate electricity that was previously generated from coal at the same total cost of $9. If, instead, the gas costs $3, producing electricity with gas instead of coal reduces the utility's costs by $1 (the distance between points A and F), and the $1 of labor and capital that is no longer needed to generate electricity can produce an additional $1 of output elsewhere in the economy.[13] Additional consumption of natural gas by firms whose demand is represented by points between A and B on the demand curve would free up smaller amounts of labor and capital, and GDP would increase by amounts between $1 and zero for each additional 1 mcf of gas used.

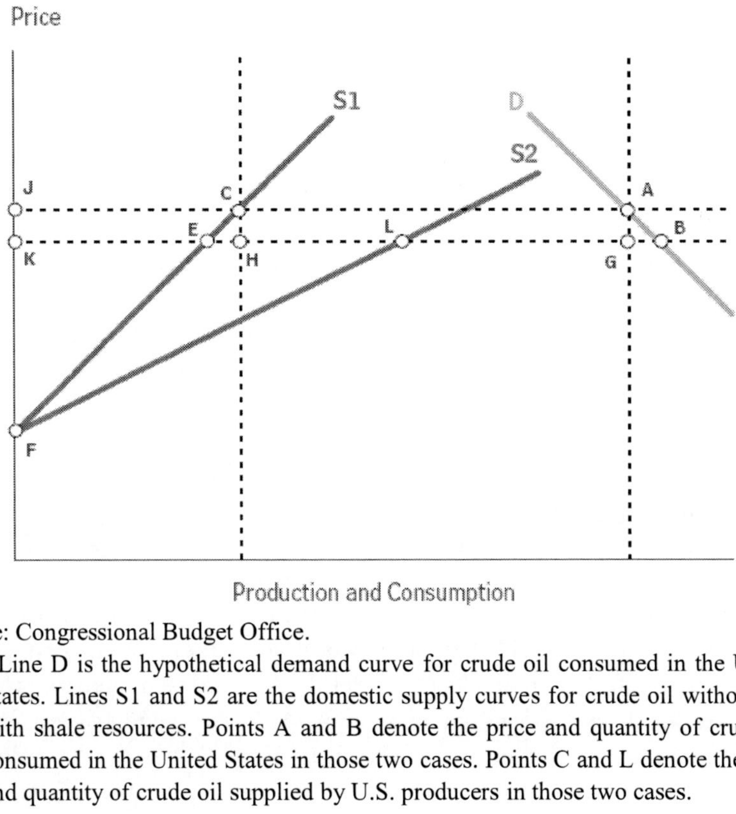

Source: Congressional Budget Office.

Note: Line D is the hypothetical demand curve for crude oil consumed in the United States. Lines S1 and S2 are the domestic supply curves for crude oil without and with shale resources. Points A and B denote the price and quantity of crude oil consumed in the United States in those two cases. Points C and L denote the price and quantity of crude oil supplied by U.S. producers in those two cases.

Figure B-7. Hypothetical Long-Run Market for Crude Oil in the United States.

Similar logic applies in the case of shifts in demand toward goods and services that are more gas-intensive. For an additional unit of a gas-intensive product, a household or firm at point A is willing to pay $4 per mcf for the natural gas that went into making the product, plus the other costs of the product, instead of spending the same total amount on other goods or services.

If natural gas costs $4 per mcf, then such a shift does not increase GDP, though it does change GDP's composition: The same total quantity of resources not used to produce the goods or services forgone is used to produce the additional unit of the gas-intensive product. If natural gas instead costs $3 per mcf to produce, the buyer's shift to the gas-intensive product reduces the total production costs of the goods purchased, and each additional 1 mcf used frees up $1 of resources that can produce additional GDP.

CBO estimated the gain in GDP by multiplying the estimated change in U.S. consumption of gas by one-half the difference between the projected price of gas and the estimated price that would prevail in the absence of shale resources. The gain in GDP is projected to be 0.1 percent in 2020 and also in 2040.

Effects of Tight Oil on the Productivity of Labor and Capital

The effects on GDP of the domestic production of tight oil differ from those of the domestic production of shale gas because oil is traded in a global market. Thus, most of the gains from greater consumption of tight oil will occur outside the United States. However, all of the gains from using labor and capital more productively to produce tight oil than they could be used for other purposes will add to U.S. GDP—by 0.2 percent of GDP in 2020 and 2040, CBO estimates.

Gains from the Increased Productivity of Labor and Capital Producing Tight Oil

The gains in GDP associated with using labor and capital to produce tight oil instead of other goods and services are illustrated in Figure B-7. The availability of tight oil shifts the supply curve of all U.S. crude oil from S1 to S2. The price of oil, established in the world oil market (which is not shown), falls from point J to point K. As a result, U.S. consumption of crude oil increases along the demand curve from A to B. The supply of conventional crude oil produced domestically falls from H to E, but the total domestic production of crude oil increases from H to L. Imports of crude oil fall, as the difference between the domestic demand for crude oil and the domestic supply narrows from the distance between A and C to the distance between B and L.

Most of the increase in GDP comes from the fact that labor and capital can be used more productively to produce tight oil than to produce other output. Using the same approach that it used when analyzing shale gas production, CBO estimated the long-term gains in GDP from the production of tight oil by multiplying its estimate of the amount of tight oil produced (which would correspond to the distance between points L and E in Figure B-7) by the difference between the price of that oil (point K) and CBO's estimate of the average break-even cost of that production (which would be equivalent to the midpoint of the supply curve between points F and L). CBO estimates that the production of tight oil will increase GDP by 0.2 percent in 2020 and 2040. In 2012 dollars, the 2040 figure is about $50 billion, which is based on projections of 4.4 million barrels per day (or about 1.6 billion barrels per year) of tight oil production in 2040; a price of roughly $140 per barrel; and an average break-even cost of about $110.

Gains in Productivity from Producing Tight Oil Instead of Conventional Oil

The gain in GDP from substituting tight oil for conventional oil that is no longer economical to produce because of the lower price of oil corresponds to the triangle bounded by points C, E, and H in Figure B-7. CBO estimated that gain in GDP by multiplying its estimate of the change in U.S. production of conventional oil by one-half the difference between the projected price of oil and the hypothetical price in the absence of shale resources. The gain is proportionally much smaller than the analogous gain for shale gas because oil is traded in a global market, which implies that the percentage impact of shale development on world oil prices is much smaller than the percentage impact on U.S. gas prices and thus that the effect on U.S. production of conventional oil is also much smaller. As a result, the effect on GDP will be very small in both 2020 and 2040, CBO projects.

Gains in Productivity from Increased Consumption of Oil

To a small degree, GDP rises as firms substitute cheaper oil for labor and capital and as goods and services produced using oil become cheaper to produce than other goods and services. The gain to GDP is reflected in the triangle bounded by points A, B, and G. Because the production of tight oil will have relatively little impact on the price of crude oil, CBO estimates that U.S. consumption of crude oil will be essentially unchanged and that the effect on GDP will be very small in both 2020 and 2040. (The majority of the gains from using more crude oil will accrue outside the United States.)

Effects of Shale Gas and Tight Oil on the Supplies of Labor and Capital

The increases in GDP associated with increased productivity would spur further increases in GDP by increasing the supplies of labor and capital. As GDP rises, households have more income to save and invest; most of the additional savings are invested domestically. That investment increases the capital stock, thus increasing the economy's productive capacity and raising GDP. In addition, higher labor productivity is reflected in higher wages, which encourage people to work and lead to an increase in the number of hours worked, likewise raising GDP. The two effects reinforce each other: A larger capital stock boosts labor productivity and wages, and an increase in the number of hours worked increases saving and investment.

CBO estimates that those indirect effects of shale development will raise GDP by 0.3 percent in 2020 and by 0.4 percent in 2040. Those estimates are based on projections of an increase of 0.1 percent in the number of hours worked in both years and of increases in the capital stock of 0.7 percent in 2020 and 0.9 percent in 2040. The projected changes in hours worked are derived from CBO's estimate that the elasticity of labor supply is 0.19 (so that a 1 percent increase in GDP per hour worked boosts the labor supply by 0.19 percent).[14] The changes in the capital stock are based on the expectation that saving and investment rise proportionally with output, so that in the long run, the percentage increase in the capital stock is equal to the percentage increase in output. CBO converted the changes in hours worked and capital stock into a change in GDP on the basis of a coefficient for labor in the production function of 0.7 and a coefficient for capital of 0.3.

End Notes

[1] For convenience, the term "shale resources" in this report includes energy supplies contained by formations of low-permeability rock that is not shale, such as limestone and fine-grained sandstone.

[2] Energy Information Administration, *Annual Energy Outlook 2014 With Projections to 2040*, DOE/EIA-0383(2014) (April 2014), http://go.usa.gov/8KyF (PDF, 12 MB).

[3] Those estimates are based on Energy Information Administration, *Technically Recoverable Shale Oil and Shale Gas Resources: An Assessment of 137 Shale Formations in 41 Countries Outside the United States* (June 2013), www.eia.gov/analysis/studies /worldshalegas/; and on Louis Sahagun, "U.S. Officials Cut Estimate of Recoverable Monterey Shale Oil by 96%," *Los Angeles Times* (May 20, 2014), http://tinyurl.com/ pnknuct.

[4] This report focuses on shale development recently enabled by the use of hydraulic fracturing and horizontal drilling. It does not consider the use of those techniques to produce "tight gas" (that is, natural gas extracted from less dense geologic formations), because such development has occurred for many years, or to enhance production from conventional oil and gas supplies. Nor does it consider kerogen shale (also known as oil shale), another kind of rock from which oil can be produced, because the oil is not extracted with hydraulic

fracturing and its generally high cost is expected to keep production low for the foreseeable future.

[5] The percentages in these two paragraphs are based on the most recently available data that distinguish shale gas from tight gas: Energy Information Administration, U.S. Crude Oil and Natural Gas Proved Reserves, 2012 (April 2014), Tables 2 and 4, www.eia.gov/naturalgas/crudeoilreserves; Energy Information Administration, *Assumptions to the Annual Energy Outlook 2013* (May 2013), Table 9.3, http://go.usa.gov/vvne; and Louis Sahagun, "U.S. Officials Cut Estimate of Recoverable Monterey Shale Oil by 96%," *Los Angeles Times* (May 20, 2014), http://tinyurl.com/pnknuct. More recent data do not distinguish shale gas from tight gas, but they do show a notable development: that the Spraberry/Wolfcamp Shale in Texas has become an important source of technically recoverable tight oil. See Energy Information Administration, *Assumptions to the Annual Energy Outlook 2014* (June 2014), Table 9.3, http://go.usa.gov/sagw.

[6] Energy Information Administration, "Natural Gas Prices" (accessed November 4, 2014), http://go.usa.gov/Kfch. Retail prices paid by final buyers of natural gas have declined less sharply, because part of the retail price of natural gas covers the cost of gas transportation and taxes—costs that do not change as the wholesale price of gas declines. In addition, partly because of differences in the percentage of retail prices that cover the cost of transportation and taxes, not all buyers have seen the same declines in retail prices: The declines have been greatest (roughly 60 percent since 2008) for large buyers, such as industrial users and electricity producers, and smallest (about 25 percent) for residential customers.

[7] Crude oil prices have declined by about a third in recent months, but not because of any sudden or dramatic increase in the availability of shale resources. Other factors, such as a rapid increase in Libyan production and a slowdown of consumption in Europe and Asia, have had a greater influence. A sustained reduction in crude oil prices would reduce U.S. production of both conventional oil and tight oil in the near term, but the degree to which the recent weakness in oil prices will persist is unclear. All else being equal, lower trajectories for the price of oil or the production of tight oil through 2020 or 2040 would reduce CBO's estimates of the economic and budgetary effects of tight oil in those years.

[8] Energy Information Administration, *Annual Energy Outlook 2014 With Projections to 2040*, DOE/EIA-0383(2014) (April 2014), http://go.usa.gov/8KyF (PDF, 12 MB).

[9] For a broader discussion of geographic price differences in world energy markets, see Congressional Budget Office, *Energy Security in the United States* (May 2012), www.cbo.gov/publication/ 43012.

[10] Energy Information Administration, *Annual Energy Outlook 2014 With Projections to 2040*, DOE/EIA-0383(2014) (April 2014), http://go.usa.gov/8KyF (PDF, 12 MB). The "low-resource" scenario includes reductions in tight gas resources as well as in tight oil and shale gas resources. Economically viable shale gas is much more plentiful than tight gas, however, and accounts for about four-fifths of the total 2040 difference in gas production between that scenario and EIA's baseline scenario.

[11] Ibid.

[12] The President is authorized to approve exports of crude oil that are in the national interest. With a few exceptions, such approval takes the form of a license from the Commerce Department's Bureau of Industry and Security. That bureau's policy is to approve certain categories of export applications—the most important category being exports to Canada for consumption or use there—and to review other applications on a case-by-case basis.

[13] Energy Information Administration, "Petroleum & Other Liquids—Exports by Destination" (accessed December 8, 2014), http://go.usa.gov/8Nvx.

[14] ICF International, *The Impacts of U.S. Crude Oil Exports on Domestic Crude Production, GDP, Employment, Trade, and Consumer Costs* (submitted to the American Petroleum Institute, March 2014), http://tinyurl.com/nnr8hxg.

[15] A more recent study yielded similar findings: In its baseline projections, lifting the ban on U.S. exports of crude oil raised those exports by no more than 1.8 million barrels per day and world crude oil production by no more than 300,000 barrels per day. See Robert Baron and others, *Economic Benefits of Lifting the Crude Oil Export Ban* (submitted by NERA Economic Consulting to the Brookings Institution, September 2014), www.nera.com/67_8673.htm.

[16] Ibid.

[17] For example, for more than three years, the price of West Texas Intermediate (WTI)—a domestically produced light crude whose price is used as a benchmark for the prices of other U.S. crude oils—has been lower than the price of Brent, a North Sea oil that is broadly representative of other world crude supplies, despite the fact that WTI is higher-quality and usually slightly more expensive. Without the ban on U.S. crude exports, the relationship between those prices would more closely reflect the historical pattern, so that the price of WTI would rise relative to that of Brent.

[18] Energy Information Administration, "U.S. Natural Gas Exports and Re-Exports by Country" (accessed November 4, 2014), http://go.usa.gov/NfKF, and "U.S. Natural Gas Imports by Country" (accessed November 4, 2014), http://go.usa.gov/Nf8B. A major reason that the United States both exports gas to Canada and imports gas from it is that some of the imports are reexported to Canada. Because it has limited pipeline infrastructure to move gas from its western regions, where most of its gas is produced, to its east, Canada serves its eastern demand by exporting gas to the western United States and importing it from the eastern United States.

[19] The United States has not been a net pipeline exporter since the 1950s. By 2040, EIA expects pipeline exports to triple and imports to decline by 30 percent. See Energy Information Administration, *Annual Energy Outlook 2014 With Projections to 2040*, DOE/EIA-0383(2014) (April 2014), http://go.usa.gov/8KyF (PDF, 12 MB).

[20] Only one LNG export facility is currently operating in the United States, and it is scheduled to export less than 0.1 Tcf of LNG over the next two years. See ConocoPhillips, "Kenai LNG Exports" (accessed August 25, 2014), http://tinyurl.com/o36tdo7.

[21] World Bank, "World Bank Commodities Price Data (The Pink Sheet)" (December 2, 2014), http://tinyurl.com/qjbfqmf (PDF, 233 KB).

[22] Department of Energy, "Summary of LNG Export Applications of the Lower 48 States" (accessed November 4, 2014), http://go.usa.gov/KfYj.

[23] W. David Montgomery and others, *Macroeconomic Impacts of LNG Exports From the United States* (submitted by NERA Economic Consulting to the Department of Energy, December 2012), http://go.usa.gov/KfGd (PDF, 4 MB).

[24] Robert Baron and others, *Updated Macroeconomic Impacts of LNG Exports From the United States* (submitted by NERA Economic Consulting to Cheniere Energy, Inc., March 2014), http://tinyurl.com/p5vcjl9.

[25] Energy Information Administration, *Annual Energy Outlook 2014 With Projections to 2040*, DOE/EIA-0383(2014) (April 2014), http://go.usa.gov/8KyF (PDF, 12 MB).

[26] CBO's calculation combined information from two sources: projections of future gas prices and production levels from EIA's *Annual Energy Outlook 2014 With Projections to 2040*;

and estimates of the price sensitivity of gas production and of the relationship between gas production and LNG exports from Robert Baron and others, *Updated Macroeconomic Impacts of LNG Exports From the United States.*

[27] Michael Levi, *A Strategy for U.S. Natural Gas Exports*, Discussion Paper 2012-04 (Brookings Institution, June 2012), http://tinyurl.com/nsxo7zo. The study's estimates were based on EIA's Residential Energy Consumption Survey for 2005, the latest information available at the time. Using data now available for 2009 does not materially affect those average costs. See Energy Information Administration, "Residential Energy Consumption Survey" (2009), Table CE2.1, http://go.usa.gov/8Kvz (XLS, 52 KB).

[28] Other policy considerations not discussed here include the effects of extending the benefits of trade in natural gas to countries that do not agree to certain provisions that generally accompany free-trade agreements (such as safer conditions for workers and environmental protection) and the possibility that some actions that the United States might take to constrain LNG exports could prompt international challenges under the rules of the World Trade Organization.

[29] Energy Information Administration, *Annual Energy Outlook 2012 With Projections to 2035*, DOE/EIA-0383(2012) (June 2012), http://go.usa.gov/7dhz.

[30] Energy Information Administration, *Annual Energy Outlook 2014 With Projections to 2040*, DOE/EIA-0383(2014) (April 2014), http://go.usa.gov/8KyF (PDF, 12 MB).

[31] James L. Coleman and others, *Assessment of Undiscovered Oil and Gas Resources of the Devonian Marcellus Shale of the Appalachian Basin Province, 2011*, FS 2011-3092 (U.S. Geological Survey, August 2011), http://pubs.usgs.gov/fs/2011/3092/.

[32] Energy Information Administration, *Annual Energy Outlook 2012 With Projections to 2035*, DOE/EIA-0383(2012) (June 2012), http://go.usa.gov/7dhz.

[33] Louis Sahagun, "U.S. Officials Cut Estimate of Recoverable Monterey Shale Oil by 96%," *Los Angeles Times* (May 20, 2014), http://tinyurl.com/pnknuct.

[34] J. David Hughes, *Drill, Baby, Drill: Can Unconventional Fuels Usher in a New Era of Energy Abundance?* (Post Carbon Institute, February 2013), www.postcarbon.org/drill-baby-drill/.

[35] Consumers who want to purchase gas or oil do not specifically demand shale gas or tight oil. Here, "demand for shale energy" at a given price refers to the excess of demand for the fuel at that price above the amount of the fuel supplied from conventional resources.

[36] Researchers recently concluded that a highly productive gas well in the Barnett Shale required a price of about $3 per million British thermal units (mmBtu) of gas to generate a 10 percent rate of return in the absence of natural gas plant liquids, but only about 50 cents per mmBtu if such liquids were present. See Peter Behr, "Barnett Shale Has Surprisingly More to Give, Texas Researchers Find," *EnergyWire* (September 25, 2013), www.eenews.net/energywire/stories/1059987786.

[37] CBO calculated that increase with data from the Bureau of Economic Analysis, "Fixed Assets Accounts Tables," Table 3.7ESI (accessed September 20, 2014), http://go.usa.gov/vwCC.

[38] Methanex Corporation, *Annual Report 2013* (Methanex, March 2014), http://tinyurl.com/lwhxkma (PDF, 601 KB).

[39] According to one report, the effect of hydraulic fracturing on investment by energy-intensive industries has been small so far but could grow in the future; see Jan Hatzius and others, *Is the Economy Gaining "Fracktion?"* US Economics Analyst 13/42 (Goldman Sachs, October 2013). Another has found that the fall in the price of natural gas since 2006 is associated with a 2 percent to 3 percent increase in activity for the entire manufacturing sector; see William R. Melick, *The Energy Boom and Manufacturing in the United States*,

International Finance Discussion Papers 1108 (Board of Governors of the Federal Reserve System, June 2014), http://go.usa.gov/vvDW (PDF, 672 KB).

40 That CBO estimate is based on Energy Information Administration, *Annual Energy Outlook 2014 With Projections to 2040*, DOE/EIA-0383(2014) (April 2014), http://go.usa.gov/8KyF (PDF, 12 MB).

41 Those estimates assume no restrictions on exports of LNG in 2020 and beyond. If the difference between domestic and overseas gas prices increased demand for U.S. exports of LNG, but those exports were constrained because federal permits had not been issued, the increases in GDP would be lower. Such a constraint would keep domestic LNG prices lower than they would be otherwise, which would benefit domestic businesses and households that used gas; however, those benefits would not fully offset the loss to gas producers. See W. David Montgomery and others, *Macroeconomic Impacts of LNG Exports From the United States* (submitted by NERA Economic Consulting to the Department of Energy, December 2012), http://go.usa.gov/KfGd (PDF, 4 MB).

42 Some researchers have estimated that shale resources will have a much larger impact on the total amount of labor and capital used in the economy in 2020, resulting in a much larger impact on GDP. For example, one report estimates that shale energy could add a net 1.7 million permanent jobs by 2020 and boost GDP by 2 percent to 4 percent; see Susan Lund and others, *Game Changers: Five Opportunities for US Growth and Renewal* (McKinsey & Company, July 2013), http://tinyurl.com/mazev4d. Another report estimates that new energy supply may create 2.7 million to 3.6 million jobs by 2020, on net, and boost GDP by 2 percent to 3 percent; see Edward L. Morse and others, *Energy 2020: North America, the New Middle East?* (Citigroup, March 2012), http://tinyurl.com/mo7k7dt. Those researchers' estimates of net jobs created are much higher than CBO's. The difference probably arises because the other researchers think that labor supply responds more strongly to increases in wages; that in 2020, the economy will still not be producing its maximum sustainable level of output (so underused labor could still be drawn into shale development without reducing the labor available to other industries); or both. For a detailed discussion of CBO's estimating approach, see Appendix B.

43 That average cost will rise because more costly resources will be profitable to develop as natural gas prices rise. The projection that it will rise more slowly than natural gas prices is consistent with EIA projections that shale gas will continue to grow as a share of overall U.S. gas production.

44 The larger initial departure from baseline prices—35 percent, compared with 25 percent in the first scenario—is consistent with EIA's recent price forecasts, which in turn reflect market expectations (shown in futures prices and trading prices for options contracts) that near-term prices have more potential to be higher than expected than to be lower than expected. See Energy Information Administration, *Short-Term Energy Outlook* (November 2014), www.eia.gov/forecasts/steo/outlook.cfm.

45 Congressional Budget Office, *Updated Budget Projections: 2014 to 2024* (April 2014), www.cbo.gov/publication/45229, and *The 2014 Long-Term Budget Outlook* (July 2014), www.cbo.gov/ publication/45471.

46 That figure is based on Congressional Budget Office, *An Update to the Budget and Economic Outlook: 2014 to 2024* (August 2014), www.cbo.gov/publication/45653. The 2015–2024 period is CBO's standard 10-year projection period.

47 Energy Information Administration, *Annual Energy Outlook 2012 With Projections to 2035*, DOE/EIA-0383(2012) (June 2012), p. 57, http://go.usa.gov/7dhz; David W. Houseknecht and others, *Assessment of Potential Oil and Gas Resources in Source Rocks of the Alaska*

North Slope, 2012, FS-2012-3013 (U.S. Geological Survey, February 2012), http://pubs.usgs.gov/fs/2012/3013/; and David W. Houseknecht, *Assessment of Potential Oil and Gas Resources in Source Rocks (Shale) of the Alaska North Slope 2012— Overview of Geology and Results* (U.S. Geological Survey, 2012), pp. 5, 7–9, http://go.usa.gov/KfeH (PDF, 4 MB).

[48] Total royalties from those regions will be higher because of the production of tight gas, which is not addressed in this report.

[49] Various proposals have been made over the years for the federal government to increase the royalties that it receives by expanding access to energy resources on federally owned land. For more information, see Congressional Budget Office, *Potential Budgetary Effects of Immediately Opening Most Federal Lands to Oil and Gas Leasing* (August 2012), www.cbo.gov/publication/43527.

End Notes for Appendix A

[1] The environmental effects discussed here are those that are associated particularly with shale development; effects that involve oil or gas more generally, such as leaks from pipelines, are not discussed. Concerns about local air pollution are also not discussed, because potential policies to address those concerns are similar to potential policies to address greenhouse gas emissions, which are discussed below.

[2] Agriculture and thermoelectric power generation each account for about 40 percent of freshwater withdrawals nationwide; the rest is used by residences, businesses, and industry (including shale development). See Molly A. Maupin and others, *Estimated Use of Water in the United States in 2010*, Circular 1405 (U.S. Geological Survey, November 2014), http://pubs.usgs.gov/circ/1405/.

[3] Texas Water Development Board, "Historical Water Use Estimates," county table for 2010 (accessed January 2014), http://go.usa.gov/KftG; Jean-Philippe Nicot and others, *Oil and Gas Water Use in Texas: Update to the 2011 Mining Water Use Report* (prepared by the Bureau of Economic Geology, University of Texas at Austin, for the Texas Oil and Gas Association, September 2012), http://go.usa.gov/KfzC (PDF, 3 MB); and Jean-Philippe Nicot, Bureau of Economic Geology, University of Texas at Austin, personal communication (January 8, 2014).

[4] See Brian D. Lutz, Aurana N. Lewis, and Martin W. Doyle, "Generation, Transport, and Disposal of Wastewater Associated With Marcellus Shale Gas Development," *Water Resources Research*, vol. 49, no. 2 (February 2013), pp. 647–656, http:// tinyurl.com/o9moyxc; and Matthew E. Mantell, "Deep Shale Natural Gas and Water Use, Part Two: Abundant, Affordable, and Still Water Efficient" (paper presented at the 2010 Ground Water Protection Council Annual Forum, Pittsburgh, Pa., September 27–29, 2010), p. 9, http://tinyurl.com/qxk2djc (PDF, 877 KB).

[5] The federal government does have some influence on water allocations. See Congressional Budget Office, *How Federal Policies Affect the Allocation of Water* (August 2006), www.cbo.gov/publication/ 18035.

[6] See Nathan Richardson and others, *The State of State Shale Gas Regulation* (Resources for the Future, June 2013), http://tinyurl.com/kwbt7l4 (PDF, 5 MB).

[7] FracFocus, "What Chemicals Are Used" (accessed December 4, 2014), http://tinyurl.com/44m94y2; Environmental Protection Agency, "Drinking Water

Contaminants" (accessed December 4, 2014), http://water.epa.gov/drink/contaminants /index.cfm; and Environmental Protection Agency, "Contaminant Candidate List 3" (accessed December 4, 2014), http://go.usa.gov/KG3m. The existing limits do cover four chemicals (benzene, ethyl-benzene, toluene, and xylene) that are less commonly present in fracturing fluid but that may be present in fracturing fluid containing diesel, which is used at perhaps 2 percent of wells; see Mike Soraghan, "Hydraulic Fracturing: Diesel Still Used to 'Frack' Wells, FracFocus Data Show," *EnergyWire* (August 17, 2012), http://tinyurl.com/puduv5m.

[8] When the report is published, it will be available at www2.epa.gov/hfstudy.

[9] Energy Information Administration, "Voluntary Reporting of Greenhouse Gases Program Fuel Emission Coefficients" (January 31, 2011), Table 1, www.eia.gov/oiaf/1605 /coefficients.html.

[10] For an overview of the measurement of emissions from natural gas systems, see A. R. Brandt and others, "Methane Leaks From North American Natural Gas Systems," *Science*, vol. 343, no. 6172 (February 14, 2014), pp. 733–735, http://tinyurl.com/ lfbaay6.

[11] Congressional Budget Office, *Energy Security in the United States* (May 2012), www.cbo.gov/publication/43012.

[12] See Intergovernmental Panel on Climate Change, *Climate Change 2014: Synthesis Report* (IPCC, 2014), p. 100, Box 3.2, Table 1, http://www.ipcc.ch/report/ar5/syr/. The impact of methane in the short term is of concern because near-term global warming may trigger a rapid, nonlinear shift from one climate state to another without the possibility of reversal; for example, see Dave Levitan, "Quick-Change Planet: Do Global Climate Tipping Points Exist?" *Scientific American* (March 25, 2013), http://tinyurl.com/ kbf247y.

[13] Environmental Protection Agency, "Summary of Requirements for Processes and Equipment at Natural Gas Well Sites" (accessed December 4, 2014), http://go.usa.gov/KGYe (PDF, 412 KB), and *Proposed New Source Performance Standards and Amendments to the National Emissions Standards for Hazardous Air Pollutants for the Oil and Natural Gas Industry* (July 2011), pp. 3–6, http://go.usa.gov/KGgH (PDF, 2 MB).

[14] For instance, one 2012 assessment assumed that on a national basis, 70 percent of the methane emissions associated with extraction were captured, 15 percent were flared, and 15 percent were released into the atmosphere. The Environmental Protection Agency, by contrast, assumed that half of those emissions were flared and half released. See Francis O'Sullivan and Sergey Paltsev, "Shale Gas Production: Potential Versus Actual Greenhouse Gas Emissions," *Environmental Research Letters*, vol. 7, no. 4 (November 26, 2012), http://iopscience.iop.org/1748-9326/7/4/044030/; and Environmental Protection Agency, *Inventory of U.S. Greenhouse Gas Emissions and Sinks: 1990–2011*, 430-R-13-001 (April 2013), pp. 3-61 and 3-62, http://go.usa.gov/KGTT (12 MB).

[15] The requirements, issued under the authority of the Clean Air Act, targeted emissions not of greenhouse gases but of volatile organic compounds and toxic air pollutants.

[16] A recent study of the methane emissions from 27 hydraulically fractured wells of companies that voluntarily participated in the study found that two-thirds of the wells—generally, those with the largest potential methane emissions—captured or controlled methane produced during the initial phases of extraction, probably in part because of the new and emerging regulatory requirements. The remaining one-third released methane into the atmosphere, but those wells had much lower emissions potential, on average. See David T. Allen and others, "Measurements of Methane Emissions at Natural Gas Production Sites in the United States," *Proceedings of the National Academy of Sciences*, vol. 110, no. 44 (October 29, 2013), pp.17768–17773, www.pnas.org/ content/early/2013/09/10/1304880110.

[17] National Energy Technology Laboratory, *Environmental Impacts of Unconventional Natural Gas Development and Production*, DOE/ NETL-2014/1651 (May 2014), pp. 39–56, http://go.usa.gov/ vvXh (PDF, 3.1 MB).

[18] See Francis O'Sullivan and Sergey Paltsev, "Shale Gas Production: Potential Versus Actual Greenhouse Gas Emissions," *Environmental Research Letters*, vol. 7, no. 4 (November 26, 2012), http://tinyurl.com/l2r8tcn.

[19] For more on environmental policy in a federal system of government, see Congressional Budget Office, *Federalism and Environmental Protection: Case Studies for Drinking Water and Ground-Level Ozone* (November 1997), www.cbo.gov/ publication/10546.

End Notes for Appendix B

[1] In this appendix, references to tight oil include not only crude oil that is extracted from shale by means of hydraulic fracturing but also CBO's estimate of the portion of the production of natural gas plant liquids—forms of natural gas that substitute for certain petroleum products—that is produced by hydraulic fracturing. Increased production of shale gas sometimes causes more natural gas plant liquids to be produced, increasing supplies of liquid fuels.

[2] Energy Information Administration, *Annual Energy Outlook 2014 With Projections to 2040*, DOE/EIA-0383(2014) (April 2014), http://go.usa.gov/8KyF (PDF, 12 MB). The estimates are based on changes in the consumption of gas or of total liquid fuels relative to changes in the production of shale gas or tight oil between EIA's "low-resource scenario" and its baseline case. CBO's estimate reflects the five-year average of those ratios from 2036 to 2040 to control for any year-to-year variability and to reflect longer-term conditions stemming from the historical absence of shale resources (rather than the near-term effects that would result from a sudden increase or decrease in the availability of shale resources).

[3] Although sensitivities at a moment in time are not directly comparable with changes that occur over time, CBO's estimated effects are qualitatively consistent with the fact that observed consumption levels have not grown as fast as shale energy production, in part because of decreases in conventional production. For instance, from 2001 to 2013, U.S. shale gas production increased by 9 trillion cubic feet (Tcf) per year, conventional gas production fell by 4.5 Tcf, and gas consumption increased by 3.5 Tcf. See Energy Information Administration, *Annual Energy Outlook 2014 With Projections to 2040*, DOE/EIA-0383(2014) (April 2014), http://go.usa.gov/8KyF (PDF, 12 MB).

[4] The estimate of 4.8 million barrels per day in 2014 consists of EIA's projected tight oil production—about 4.1 million barrels per day—and 0.7 million of the 2.5 million barrels per day that EIA projects for natural gas plant liquids. The remaining 1.8 million barrels per day of natural gas plant liquids correspond to the average production of those liquids from 2006 to 2009, just before the boom in shale gas and tight oil production. See EIA's *Annual Energy Outlook* for 2014 and for 2006 through 2009.

[5] In CBO's baseline projection, in which shale resources are assumed to be available, 29.1 Tcf of gas are expected to be produced in the United States during 2020 at a price of $4.40 per million British thermal units (mmBtu). Because there are roughly 1.03 mmBtu in each thousand cubic feet (mcf) of gas and 1 Tcf equals 1 billion mcf, sales of natural gas will total about $130 billion (29.1 multiplied by 1.03 multiplied by $4.40 multiplied by 1 billion). With total liquid fuel production expected to be 14.5 million barrels per day and an

oil price of $97 per barrel (assumed to be the same for nonpetroleum fuels), sales of liquid fuels are expected to total about $1.4 billion per day (14.5 million multiplied by $97), or $515 billion per year. In the absence of shale resources, CBO projects, 2020 gas and oil prices would be $6.90 per mmBtu and $103 per barrel; U.S. gas and oil production would be 16.7 Tcf and 9.9 million barrels per day; and total spending on natural gas and liquid fuels would be about $495 billion.

[6] The labor and capital used to produce shale energy include what is used to produce goods and services subsequently employed in shale development—for example, the labor and capital used to produce the concrete that, in turn, provides the casing of a new well.

[7] International Energy Agency, *Resources to Reserves 2013: Oil, Gas and Coal Technologies for the Energy Markets of the Future* (IEA, 2013), www.iea.org/w/bookshop/add.aspx?id=447.

[8] See the low-resource scenario in Energy Information Administration, *Annual Energy Outlook 2014 With Projections to 2040*, DOE/EIA-0383(2014) (April 2014), http://go.usa.gov/8KyF (PDF, 12 MB). The actual production of shale gas and tight oil in that scenario falls by less than 50 percent because the effects of the decline take some time to materialize and because higher market prices promote the development of additional wells.

[9] EIA also analyzed a high-resource scenario in which shale gas and tight oil wells were more productive than they were in the agency's baseline projection. CBO did not use that scenario to calculate any of its range of shale production quantities because the scenario included changes to *conventional* oil and gas supplies that were not included in EIA's low-resource scenario.

[10] Because of rounding, the total change in real GDP is slightly less than the sum of the component changes.

[11] The market that determines the domestic price of natural gas includes supply and demand in Canada and Mexico. To focus on the effects on GDP in the United States, the reader should interpret the supply curves, S1 and S2, as representing domestic supply, and the demand curve, D, as representing demand net of supplies from Canada and Mexico.

[12] The area enclosed by points A, B, and F also includes the gain to foreign users that are able to consume more natural gas. In calculating the gains to U.S. GDP, CBO considered only the changes in domestic consumption.

[13] The utility itself might not reduce its use of labor and capital, but those resources would be freed up elsewhere in the economy— particularly in coal production and related activities. Note that the effect on the *composition* of GDP exceeds the effect on its size: In this hypothetical situation, the output of the natural gas industry increases by $3 and GDP increases by $1.

[14] Congressional Budget Office, *How the Supply of Labor Responds to Changes in Fiscal Policy* (October 2012), www.cbo.gov/ publication/43674. CBO's labor supply elasticity is the sum of its estimates of the substitution elasticity (how much an increase in wages increases the amount of labor supplied because working becomes more valuable relative to other uses of people's time) and of the income elasticity (how much an increase in wages allows people to work fewer hours while maintaining their standard of living).

In: Producing Oil and Natural Gas … ISBN: 978-1-63482-126-1
Editor: Joelle Bolton

Chapter 2

U.S. Shale Gas Development: Production, Infrastructure, and Market Issues*

Michael Ratner

Summary

The United States is in the process of discovering the true extent of its unconventional natural gas resources primarily derived from shale, with estimates rising every year. Natural gas has become an increasingly available, low-priced commodity, and U.S. natural gas is in demand at home and abroad. The price of natural gas relative to coal has led to a trend of switching from coal to gas for electric generation.

Estimates of undiscovered technically recoverable natural gas have multiplied in recent years, rewriting the map on natural gas production. Pennsylvania is newly one of the top four gas-producing states (since 2011), with Texas holding the top spot. This rise has necessitated pipeline network expansions; however, progress has been slow, leading to constraints and price differentials. Appalachian natural gas has historically been priced higher than that from the Gulf Coast's Henry Hub due to lower availability, but increased production in Pennsylvania's Marcellus Shale resource has reversed this differential.

* This is an edited, reformatted and augmented version of a Congressional Research Service publication R43636, prepared for Members and Committees of Congress dated July 14, 2014.

The earning potential for liquefied natural gas (LNG) exportation overseas, particularly in high-demand areas such as Asia, has led to an increase in applications to construct U.S. export terminals. The prospect of increased LNG exports raises three main issues. First, exporting may raise domestic prices just as domestic natural gas usage is beginning to expand. Second, an influx of LNG into the global market may create new geopolitical concerns. Natural gas production in the United States has prompted other countries to attempt to produce from shale deposits. Third, onshore export terminals may be built in ecologically sensitive areas with the alternative being offshore terminals. Additionally, an increased reliance on LNG may shift focus away from exploring renewable energy sources.

The increase in U.S. natural gas production also affects the domestic market. The percent of new housing constructed to be heated by natural gas has become almost equal to the percent heated by electricity. This increase has been due largely to economic and environmental considerations. In the transportation sector, compressed natural gas (CNG) and LNG are beginning to displace traditional liquid fuels; some companies are producing vehicles that run on various forms of natural gas. However, the main obstacles are the cost of the new technology required and that of reformatting fueling stations.

UNITED STATES: IMPETUS FOR GLOBAL CHANGES

With shale gas resources in North America and around the world continuing to be identified, evaluated, and developed, there is an increased expectation by many in academia, government, and industry that there is now resource abundance. The United States has led this trend, and the overall U.S. natural gas supply likely will expand in the near term and beyond. The U.S. Energy Information Administration (EIA), in the *Annual Energy Outlook 2014* projections, reflects this advancing pace in natural gas production. The United States is facing increasing industrial and power demands for natural gas, as well as possibly becoming a net liquefied natural gas (LNG) exporter and overall gas exporting country by 2016 and 2018, respectively.[1] At the same time, *Figure 1* illustrates how prices for the commodity are continuing to trend near historical lows. Current expectations are for those levels to continue in the near and mid-terms.

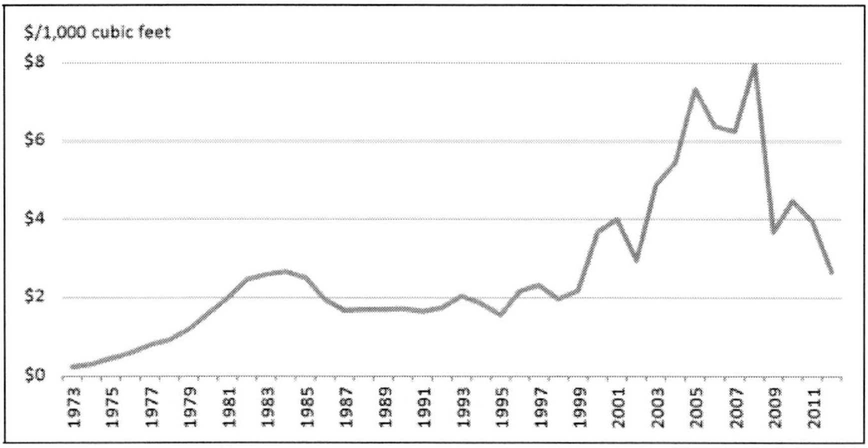

Source: U.S. Energy Information Administration, Natural Gas Prices database, http://www.eia.gov/dnav/ng/hist/n9190us3a.htm.

Notes: Data are annual average prices in nominal dollars.

Figure 1. U.S. Annual Wellhead Natural Gas Prices 1973-2012.

While low prices are generally good news for consumers in the industrial, power generation, commercial, and residential sectors, they do cause a number of closely related technical and market impacts worth further research and analysis. Recent changes in thermal coal use in the United States, illustrated in *Figure 2*, impacted by switching from coal to natural gas, are substantial and will likely continue to trend downward due to overall market influences of natural gas prices relative to coal, along with current and pending federal and state policy.[2] Coal-to-gas switching for U.S. power generation doubled in 2012 from the same period in 2011, with gas use for power generation doubling from 3 billion cubic feet per day to 6 billion cubic feet per day.[3]

Although coal export to Asian and European electric power generation operators has risen, there has still been a net negative impact on the mining sector in several regions of the United States. Rail capacity that was used to move coal to market is now underutilized. On the other hand, some of the rail capacity is now being redeployed to move crude oil from shale deposits in the upper western United States to eastern refineries where demand for domestic West Texas Intermediate (WTI) priced crude has increased relative to higher-cost Brent priced imports from other global locations.[4] Pipeline capacity to accomplish the same goal has regulatory, right-of-way, and financial challenges, and is viewed by some as insufficiently responsive to the rapid increase in development. It is possible that rail will one day move natural gas as well.

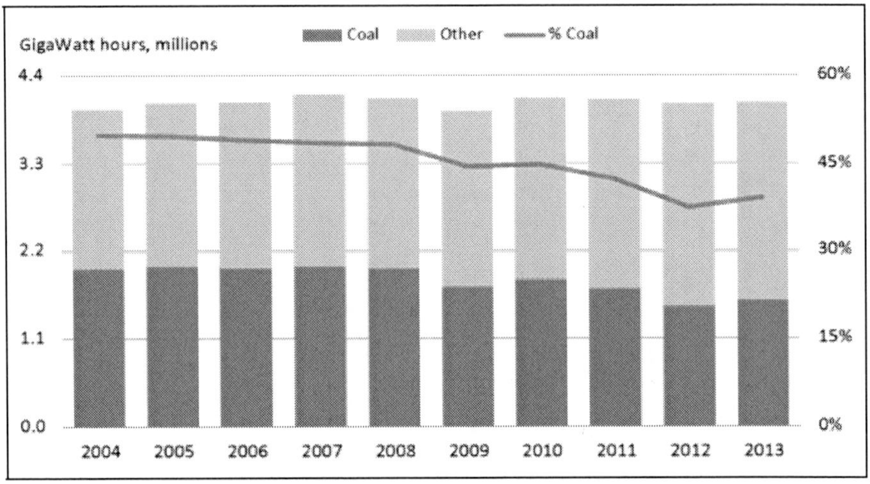

Source: U.S. Energy Information Administration, Electric Power Monthly, http://www.eia.gov/electricity/monthly/epm_table_grapher.cfm?t=epmt_1_1.
Notes: Units = GigaWatt hours (GWh).

Figure 2. U.S. Electricity Production from Coal.

The addition of a large number of U.S. and international oil and natural gas companies entering shale gas developments during the last six years, with larger capital budgets and longer investment horizons, has advanced both private and public research on natural gas extraction technologies. The presence of these companies also may increase the likelihood of other countries' developing shale formations. Field service companies are also integral in both these respects. Government agencies have traditionally conducted or funded related energy research and currently are pursuing multiagency initiatives, although federal budgetary constraints may limit the extent of these activities going forward.

NEW PRODUCTION, CHANGING LOCALES

A historical shift is taking place in how fossil energy is being sourced and utilized when compared to energy development and consumption patterns of the past four decades. Conventional thinking is that North America will now make very significant gains in the goal of achieving energy independence due to both reduced energy use through efficiencies, i.e. increases in the Corporate Average Fuel Economy (CAFE) standards, among other things, and rapidly increasing supply. The technologies and techniques being applied to develop new sources of natural gas from shale are advancing quickly as the science is evolving and the overall size of the resource is better understood. As an example, early estimates of the Marcellus Shale resource in the Appalachian basin proved somewhat conservative, and have been revised upward by the U.S. Geological Survey (USGS) from the original 2 trillion cubic feet to the current projection of 84 trillion cubic feet of undiscovered technically recoverable gas.5 Figure 3 illustrates recent shale-gas resources estimates.

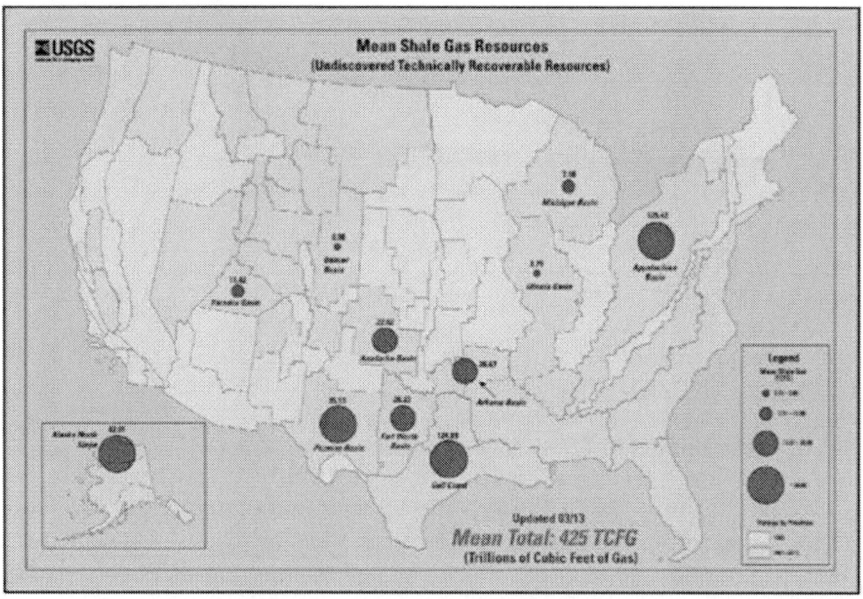

Source: U.S. Geological Survey, National Oil and Assessment 2013 Assessment Updates, http://certmapper.cr.usgs.gov/data/noga00/natl/graphic/2013/mean_shale_gas_2013_large.png.

Notes: Units = trillion cubic feet of gas (tcfg).

Figure 3. USGS Shale-Gas Resource Estimates.

Some third-party firms and academics now indicate that this shale alone could yield upwards of four times the most recent previous estimates, based on a better understanding of the geology, much of which is now based on actual drilling and the resultant data coming from those wells.[6] In the case of

Pennsylvania, more than 6,000 wells had been drilled into the Marcellus as of November 2012,[7] and in 2012, Pennsylvania produced more than 2 trillion cubic feet of natural gas (see *Figure 4*). Similar data from West Virginia's shale wells to date indicate a similar trend. In that multistate region, gas volumes are increasing even as the number of drilling rigs operating shows a slowly decreasing year-by-year trend.

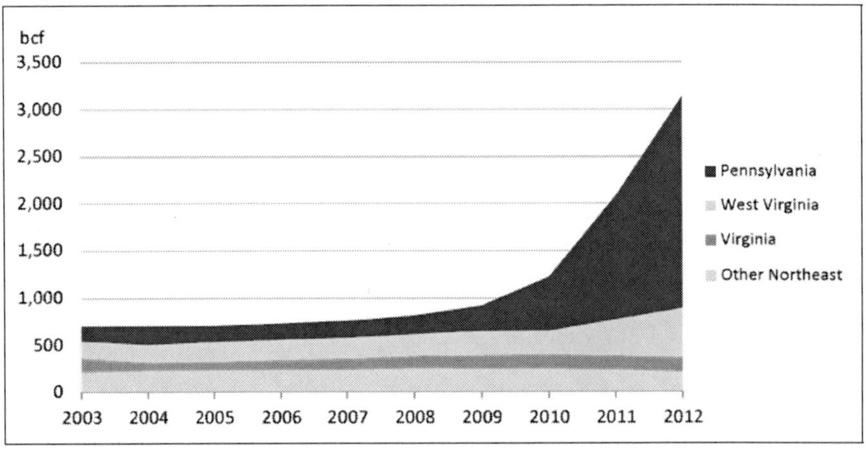

Source: U.S. Energy Information Administration, Natural Gas Gross Withdrawals and Production, http://www.eia.gov/dnav/ng/ng_prod_sum_dcu_NUS_a.htm. Notes: "Other Northeast" includes Kentucky, Maryland, New York, Ohio, and Tennessee. Units = billion cubic feet per year (bcf).

Figure 4. Annual Northeast Natural Gas Production.

Figure 5 shows that a majority of the natural gas produced in the United States comes from a small group of states, led by Texas. However, it is important to note that Pennsylvania's rise into the top four producing states is a relatively recent occurrence. Prior to 2011, Pennsylvania was producing less than 1 trillion cubic feet per year, and before then it produced less than 0.2 trillion cubic on average since the 1980s. Pennsylvania's rise in production is significant for overall U.S. supplies, but more importantly because of its proximity to the largest U.S. markets for natural gas, the Northeast.

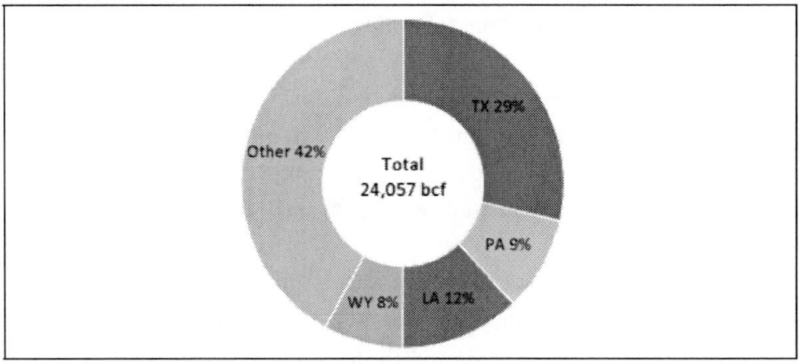

Source: U.S. Energy Information Administration, Natural Gas Gross Withdrawals and Production, http://www.eia.gov/dnav/ng/ng_prod_sum_dcu_NUS_a.htm..
Notes: Production volumes are for dry natural gas. Units = billion cubic feet (bcf).

Figure 5. Top 4 Natural Gas Producing States (2012).

Infrastructure Requirements

Pipeline capacity in the same region of the United States, both interstate transmission lines and local gathering lines, is currently in need of expansion to accommodate the potential demands of energy companies completing natural gas wells. Approximately 50% of the wells drilled in Pennsylvania do not have access to a pipeline allowing gas to reach commercial markets. This is particularly true in northeastern Pennsylvania, where the industry has a shorter history of drilling than in other regions of the United States, thus necessitating the need to construct more of this infrastructure, particularly gathering-pipeline systems. The gas market is accounting for this bottleneck in supply near large eastern cities, with a lower basis price at key citygate locations when compared to historical premium price levels.[8] In western Pennsylvania, where the natural gas also contains natural gas liquids, fractionation facilities are also an infrastructure constraint to development. Industry estimates indicate there will need to be a nationwide investment of $10 billion per year for each of the next 10 years to develop adequate capacity in gathering and transmission pipelines, along with associated compression facilities en route.[9]

Similar trends are also being seen in gas flows in other regions. The Rockies Express Pipeline (REX) built to supply gas from the Rocky Mountains to eastern U.S. markets was at 70% capacity during the winter of 2012.[10] Its unused capacity is largely accounted for due to the net surplus of

Marcellus gas produced in Pennsylvania and West Virginia. The combined transmission capacity of the historical pipelines routed from the Gulf Coast to northeastern markets was at 45% of potential volume in 2012.[11] Legacy premium gas pricing seen in northeastern U.S. markets will likely trend to western markets within five years due to increasing demand there; eastern shales are predicted to expand net surpluses in that region during the same period.[12]

West-to-east flow in trans-Canadian pipeline capacity has declined as well, partially related to eastern United States shale gas moving into eastern Canada in 2012 at a rate of 400 million cubic feet per day due to closer proximity and upgraded pipeline and compression facilities.[13] In addition, declining Canadian gas production and greater use in western Canada are contributing factors in this trend. Overall, the amount of natural gas being imported from Canada, a traditional supplier, has been dropping (see *Figure 6*), and is now less than 8% of total U.S. usage. Similarly, flow of gas to Mexico is increasing, exceeding 1.8 bcf/d in 2013.

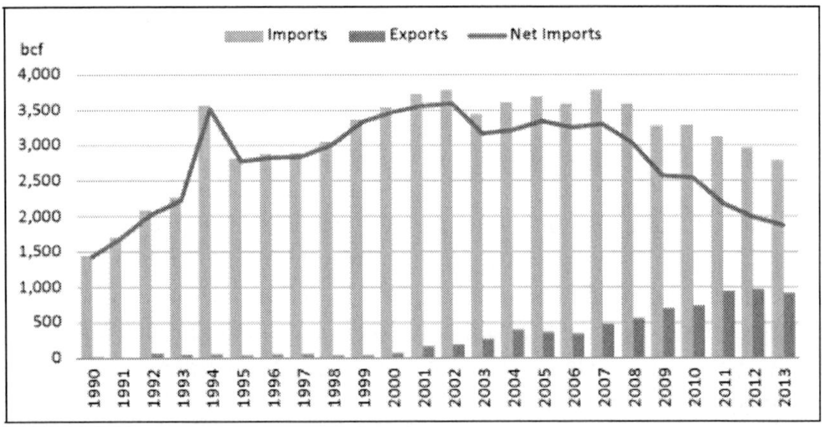

Source: U.S. Energy Information Administration, U.S. Imports by Country and U.S. Exports by Country databases, http://www.eia.gov/naturalgas/data.cfm#imports.
Notes: Units = billion cubic feet per year (bcf).

Figure 6. U.S. Natural Gas Net Imports from Canada.

Changing Price Differentials

The traditional market dynamics in the United States between supply and demand are changing and affecting how natural gas is priced. The Henry Hub in Erath, LA, has been the basis for U.S. natural gas prices, in part because of the numbers of natural gas pipelines that intersect that area. With more natural gas being produced outside of the Gulf area and the amount of natural gas passing through Henry Hub, market distortions are arising that are being reflected in regional prices. For example, the spreads between the Columbia Gas Transmission Appalachian index (TCO Appalachia)—an actively traded location for both physical and financial transactions for natural gas in southwest Pennsylvania—and the Henry Hub are illustrated in *Figure 7*. The nature of the spread is changing due mainly to growth in Marcellus production. Natural gas at the TCO Appalachia has historically been priced about $0.25 per million British thermal units above Henry Hub. However, the spread between these two points in spot markets reflects roughly parity now, and in forward markets TCO Appalachia is priced less than at the Henry Hub. This series reflects how some market participants assess the future difference between the southwest Pennsylvania and Gulf Coast natural gas market.

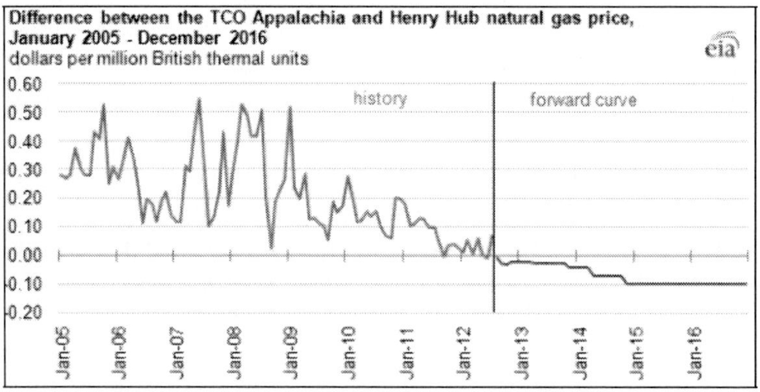

Source: U.S. Energy Information Administration, based on Bloomberg LP, Marcellus natural gas production gains affect spreads between trading points, July 24, 2012, http://www.eia.gov/todayinenergy/detail.cfm?id=7230.

Notes: The blue series shows the historical difference between the monthly average spot natural gas price at the TCO Appalachia trading point and Henry Hub trading point. The red series reflects the forward curve or slate of monthly values for the TCO Appalachia CME Group NYMEX ClearPort basis swap as of July 19, 2012. This basis swap is a financial instrument representing the difference in the value between natural gas in southwest Pennsylvania and the natural gas futures contract at the Henry Hub in the Gulf Coast for the period July 2012 through December 2016.

Figure 7. Difference Between the TCO Appalachia and Henry Hub Natural Gas Price, January 2005 Through December 2016.

Market Considerations and Responses

As the marketplace better realizes the share that shale gas production is able to add to the total volume of U.S. production, and anticipates the potential drilling successes of energy companies developing this resource, short- and mid-term prices are being held to the lower end of the historical range. This affects companies' selection of regions and locations to drill as they seek the lowest-cost wells to develop. *Figure 8* illustrates the variation in break-even prices required in different shale-gas regions. Break-even prices are a function of both geography and the composition of produced hydrocarbons (associated natural gas liquids generally reduce the break-even price by around 50%, compared to dry gas deposits, since those liquids have a high market value aside from the saleable gas). Likewise, break-even prices can potentially influence the number of wells companies choose to drill in a low commodity price environment, compelling them to promote greater use of natural gas (i.e., new transportation fueling options, increased onshore production of polyethylene, redevelopment of U.S.-based nitrogen fertilizer production, etc.), and prodding them to seek higher-priced markets in which to sell their final product. Additionally, the time for a well to pay back its cost is an important factor in development. Shale wells, which usually have a higher probability of success of finding natural gas than traditional wells in traditional formations, have a short payback time, usually within two years but sometimes within months, depending upon production rates, market prices, and other factors.

LNG Export Factors

Exports of natural gas to take advantage of the higher prices seen for the commodity in other regions of the globe are an emerging opportunity for many segments of the energy industry involved in developing shale in the United States. Prices for equal units of gas in Europe or Asia can be five to seven times the value of the natural gas priced at Henry Hub. Gas sold on the global market is commonly indexed to crude oil market prices. The Henry Hub price does not reflect the same indexing to crude oil prices, and is more regional than global in nature. The margin between the two creates a potential export opportunity for U.S. producers. As of April 2014, there were 43 applications at the U.S. Department of Energy (DOE) from firms interested in developing export terminals to sell gas offshore. Six facilities in the United States have

already received at least conditional approval to export LNG to non-free trade countries (non-FTA).[14]

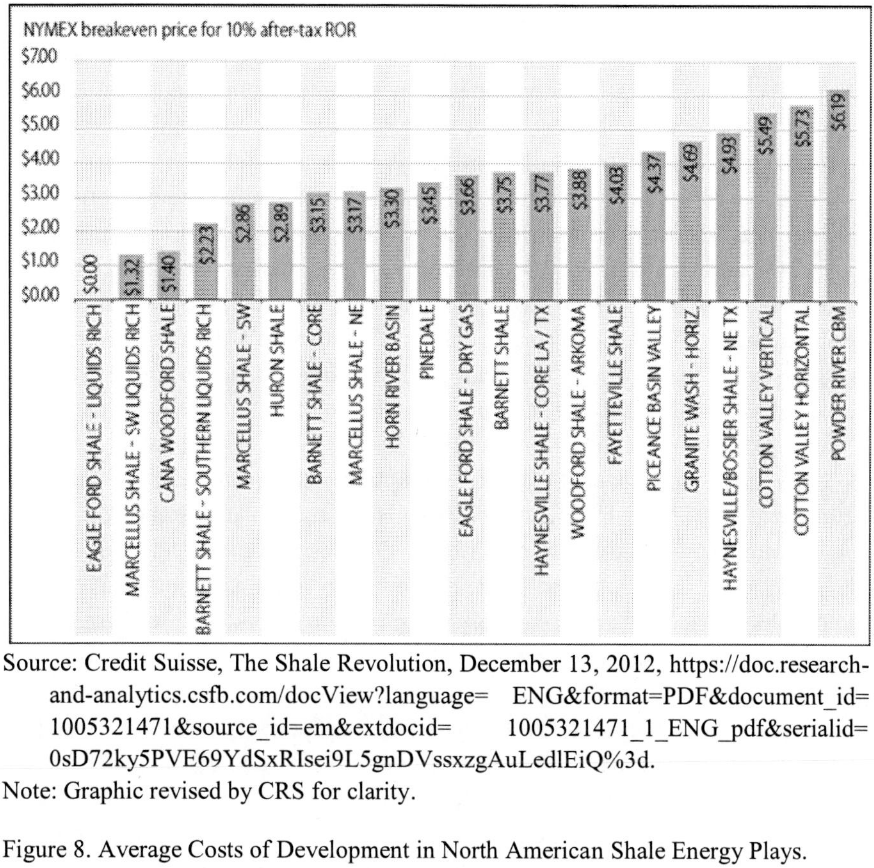

Source: Credit Suisse, The Shale Revolution, December 13, 2012, https://doc.research-and-analytics.csfb.com/docView?language= ENG&format=PDF&document_id= 1005321471&source_id=em&extdocid= 1005321471_1_ENG_pdf&serialid= 0sD72ky5PVE69YdSxRIsei9L5gnDVssxzgAuLedlEiQ%3d.

Note: Graphic revised by CRS for clarity.

Figure 8. Average Costs of Development in North American Shale Energy Plays.

Along with potential price impacts to the United States, four other primary issues could surface from the export of larger amounts of LNG.

1. Industrial Utilization. Natural gas is used as both a heating fuel and chemical feedstock for consumers in the United States and around the globe. Concern has been expressed from some voices in the industrial and utility sectors that LNG exported overseas will raise natural gas prices in the United States.[15] There is an emerging trend toward the re-shoring of manufacturing (i.e., the return of formerly offshored manufacturing to the United States), in addition to new development by both U.S. and international companies. Growing numbers of manufacturers are considering more predictable energy supplies and conditions in the United States, both in quantity and favorable pricing, as being a key decision point in where to site new or expanded capacity.[16] Recent investments by several large petrochemical companies to enhance their capacity in the Gulf Coast region, along with other locations in the United States, are examples of an emerging trend.[17] Estimates predict that construction of new petrochemical plants could exceed $80 billion in the next five years.[18] Similar capital investments are currently being made domestically in high energy-use industries such as paper, glass, and ammonia fertilizer, as well as aluminum, steel, and other primary metals production.

In a similar manner, electrical generation in the United States from coal is declining as gas utilities are being dispatched ahead of coal. The permitting period for construction of new gas-fired power plants is also shorter than for coal or nuclear units. Data show that much of this natural gas is now being sourced onshore versus offshore in the Gulf of Mexico.[19] Increasing quantities of gas from shale resources, fuel sourcing less prone to price volatility (especially during hurricane season), and lower construction costs and time frames for new combined heat-and-power gas units to replace aging coal capacity taken offline have created an uptick in demand for the increasing volumes of shale gas hitting the market. High and volatile prices for natural gas, particularly relative to coal, over the past three decades have been constraints on building gas-fired capacity. In contrast, coal has offered predictable pricing, fewer supply interruptions, and a well-established transportation infrastructure for delivery.

When new re-shoring is combined with existing and growing amounts of gas-fired power generation, these two sectors collectively have the capacity to use substantial quantities of natural gas, and accordingly would be sensitive to factors that could lead to moderate or higher energy pricing. DOE commissioned two studies as part of its public interest determination required by law for natural gas exports, one by EIA on domestic price impacts and one by NERA Economic Consulting on economic impacts. The EIA study determined a range of price increases that may happen depending upon the level and circumstances of LNG exports, and the NERA study concluded that LNG exports would yield positive net economic benefits to the U.S. economy in all scenarios. Other studies have been done by private and public institutions, but are beyond the scope of this report.

2. Geopolitics. The successes in extracting natural gas in North America, combined with emerging knowledge of the commercial value of shale geology and location, have prodded many other nations around the world to attempt to pursue similar natural gas development. Very large shale deposits with the potential for energy development have been identified in other regions of the globe.[20] Foreign governments are moving forward in matching their regulatory expertise and infrastructure needs to the emergence of this source of natural gas. Argentina, China, Britain, South Africa, and Poland have been some of the early movers on shale gas development. The pace of this new development has been highly variable to date. Reasons cited include largely unexplored geology, internal policies and politics, constraints in infrastructure, lack of a gas field service industry sector, and, in the short term, potentially limited availability of needed technology.

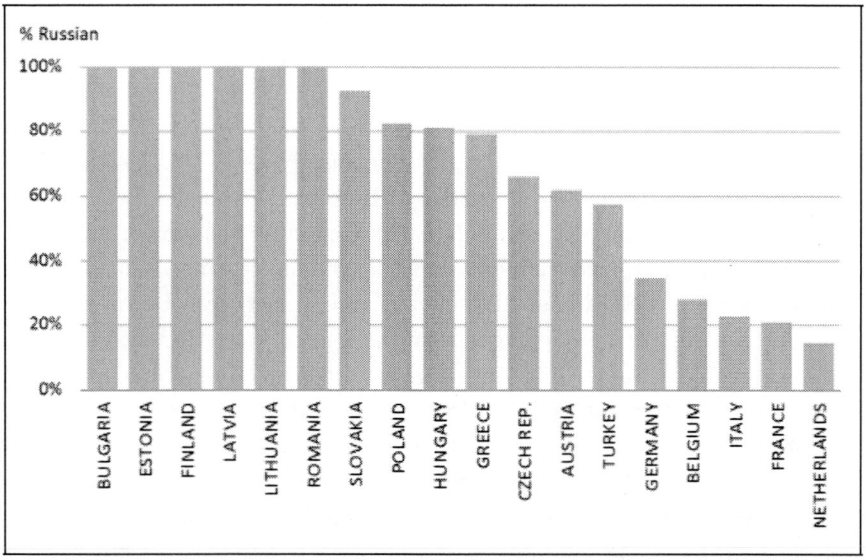

Source: BP Statistical Review of World Energy, http://www.bp.com/content/ dam/bp/pdf/statistical-review/statistical_review_of_world_energy_2013.pdf, and Eurogas, http://www.eurogas.org/uploads/media/Eurogas_Statistical_Report_ 2013.pdf.

Figure 9. European Dependence on Russian Natural Gas Imports, 2012.

Geopolitics may be influenced by the development of gas from shale in the United States and abroad. Natural gas sourced from shale is now more than one-third of the U.S. total, and is expected to increase to nearly 50% by 2035. This is putting the nation into an oversupplied situation, which in turn is pushing back onto the world market cargos of LNG that originally were developed and designated to cover declining U.S. production. Some estimates indicate that 26% of current worldwide LNG capacity was developed for the U.S. market. Strong demand for LNG is being seen from Japan post-Fukushima, South Korea, and Europe, all of which are key U.S. allies. Any potential export of U.S. LNG would likely go to these countries due to market forces and proximity. Efforts to reduce reliance on nuclear power generation in some nations, along with evolving climate policy, will generate additional demand for LNG.

Currently, the largest supplier of natural gas to many countries in Europe is the Russian entity Gazprom, which is majority state-owned. (See *Figure 9.)* Disputes between Gazprom and Ukraine over pipeline transit and supply issues have led to interruptions to Europe in gas deliveries during the critical winter heating seasons, which have led to efforts by European Union members to look at means to diversify their supply.[21] The standoff between Russia and Ukraine in 2014 has halted natural gas exports to Ukraine from Russia, but has not disrupted supplies being sent to Europe. Along with the push for diversification, expected U.S. LNG export to the European Union is also affected by the advent of shale development successes in North America; the realization of shale potential in regions of Europe; the planned dismantling of nuclear power generation, particularly in Germany; the broad increase in renewables; the Eurozone economic crisis, which is causing demand erosion; the permitting of LNG import facilities in several European Union countries; and the potential for LNG exports from the United States to reach Europe in the foreseeable future. Gazprom is also under investigation currently by European Union regulators on business practice irregularities. This is leading to pricing pressure on long-term natural gas contracts being negotiated in the near term between European governments and Gazprom. Poland, for instance, has stated its intention to double current national production, utilizing its new efforts with shale gas exploration and development, before a new contract with Gazprom is finalized in 2020.[22] And in a similar fashion in the United States, DOE, the State Department, and the Department of the Interior have been coordinating an effort to develop greater global energy security, with an emphasis on the development of natural gas resources from shale in countries where dependence on countries such as Russia and Iran is high. The State Department has also been advocating for gas transmission lines that increase the diversity of suppliers in countries where it has influence in the negotiation of pipeline rights-of-way.

3. Balance of Trade. Initially the U.S. trade balance benefitted by the rise in production of natural gas, which negated the need for much of the natural gas the country imported. Most U.S. imports of natural gas have come from Canada and since 2007 have declined 26%. Imports of LNG were on a steady rise from 1995 and peaked in 2007 when, similar to pipeline imports from

Canada, they started to decline. Between 2007 and 2013, U.S. LNG imports dropped by 87%.

4. Organized Environmental Response. Opposition from some segments of the environmental community to the export of gas is generally based on two issues.

- First, export terminals may be built in ecologically sensitive coastal areas. In some cases, newly proposed LNG export terminals may be built as expansions of underutilized import facilities (Cove Point on Maryland's Eastern Shore, for example).[23] Undeveloped land would be converted to industrial use for such projects. As an alternative approach, new technology could be used to build offshore floating terminals, which, although they have an onshore component, lessen land-based impacts.
- Second, development of LNG would promote further development of shale resources, which is counter to the effort to reduce national reliance on fossil energy of all types. Terminal construction and associated exports would be predicated on the upstream development of shale wells, particularly in states near the East and West Coasts, and on the Gulf Coast to a lesser degree due to large conventional reserves of gas already expected to be bound for market. Groups contesting new LNG applications in the United States are calling for the evaluation of upstream environmental impacts as part of environmental reviews during LNG site permitting.[24]

Related to the issues about shale development because of LNG exports, there is concern by some of the public over gas development in their communities; how to deal with the potential risks from energy extraction; and attempts to determine the benefits to them as individuals, their communities, or the nation on a larger scale. Exporting energy, particularly natural gas from shale, is drawing an adverse reaction from some. Concerns heard by elected officials in some of these communities from constituents have included experiencing the impacts of development, such as increased truck traffic, but not seeing direct benefits of development.

EXPANDING USE OF NATURAL GAS

With a growing awareness of both the size of the natural gas resource likely to be sourced from shale and the rate at which it is being produced, interest in domestic use is increasing in all sectors. The use of gas-fired residential heating in both new construction and retrofitted homes near existing gas distribution lines is at near record pace in some regions, particularly in the northeastern United States, where fuel oil has the largest use nationwide. (See *Figure 10.)* Increases are also occurring in densely populated areas such as New York City, through a project to collectively save more than $250 million annually, and largely reduce the use of certain heavy fuel oils for heating high-rise apartments in many sections of the city.[25] New York City has been phasing out certain heavy fuel oils since 2011 because of health effects from emissions. Beyond the economic value of the transition from fuel oil to gas, there are environmental impacts in the form of reduced air emissions.

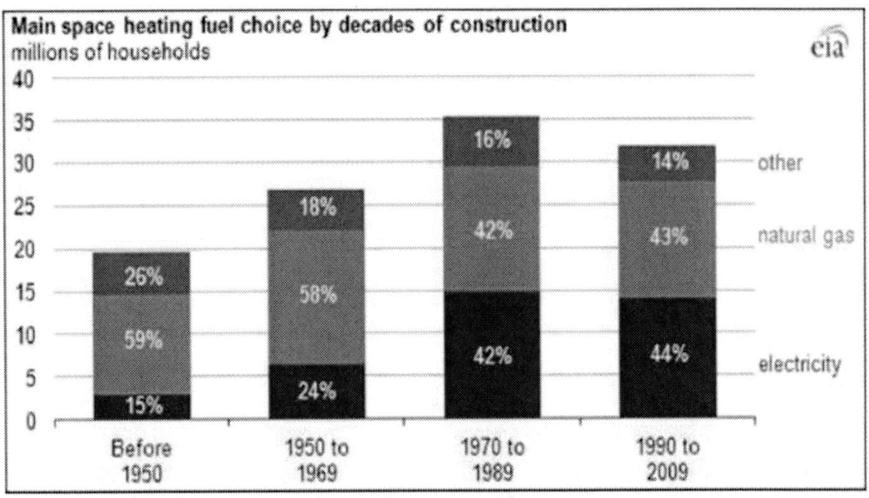

Source: U.S. Energy Information Administration, Heating fuel choice shows electricity and natural gas roughly equal in newer homes, http://www.eia.gov/todayinenergy/detail.cfm?id=7690.

Figure 10. Main Space Heating Fuel Choice by Decades of Construction.

As illustrated in *Figure 11*, a comparison of prices for natural gas and fuel oil for domestic home use shows a marked divergence.

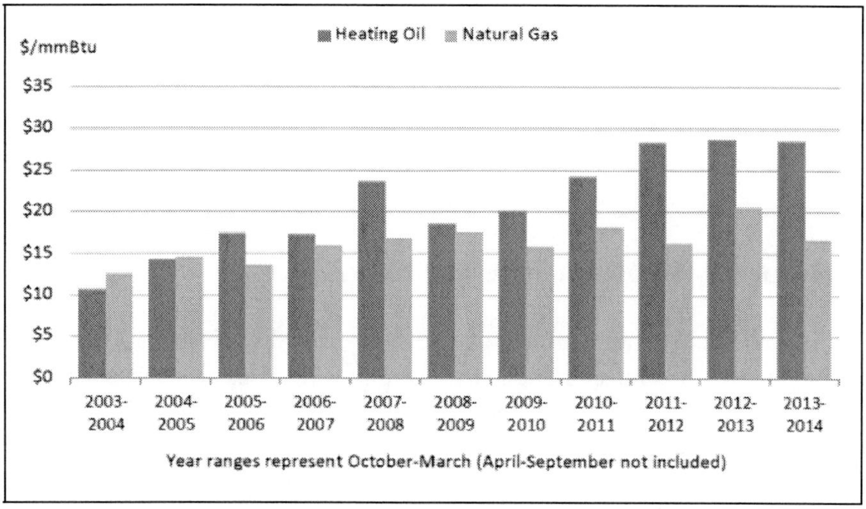

Source: U.S. Energy Information Administration, Heating oil and natural gas databases, http://www.eia.gov/dnav/pet/hist/LeafHandler.ashx?n=PET&s=M_EPD2F_PRS_NUS_DPG&f=M and http://www.eia.gov/dnav/ng/ng_pri_sum_dcu_nus_m.htm, respectively.

Notes: Northeast includes the states of Connecticut, Maine, Massachusetts, New Hampshire, Rhode Island, and Vermont. Units = dollars per million British thermal unit ($/mmBtu).

Figure 11. Winter Heating Oil and Natural Gas: Northeast United States Residential Prices.

Utility companies are increasing construction of new distribution lines for all types of customers due to the predicted favorable pricing of natural gas in the mid term based on increased supply. This includes the extension of existing lines, which can cost up to $1 million per mile in some populated locations. There is also a continued modernization of existing lines due to ongoing replacement of the aging distribution infrastructure, greater regulatory scrutiny, and the demand for increased volumes of gas. Additionally, the locations of shale gas deposits tend to differ from traditional natural gas production areas, causing infrastructure to be reconfigured.

Awareness is growing at the state and local government levels that areas not served by natural gas are becoming disadvantaged when competing for new industry to locate in their jurisdictions. This is leading to new efforts to assess the cost and feasibility of new distribution infrastructure in many regions of the United States as a means of enhancing economic development. Demand for gas trunk lines installed to a new gas-consuming industry site has the spin-off effect of making it advantageous to develop additional distribution to areas in between. Due to the geographical range of many shale plays in the United States, some existing energy-intensive industries are seeking to source shale gas directly under the footprint of their own facilities or directly from nearby producers in an effort to eliminate gas transmission costs.[26]

Transport Fuels Alternatives for Natural Gas

Another key trend in utilization is the expected displacement of traditional liquid fuels for transportation by natural gas. This is currently being accomplished through the compression of natural gas to compressed natural gas (CNG) or chilling and liquefaction to LNG, which is used in vehicles and trucks with standard engines modified to utilize natural gas. This approach necessitates the need for new on-board fuel tanks, which are large and relatively expensive, particularly when installed as retrofits on light-duty trucks or passenger cars. Several major car and truck makers in the United States are producing limited numbers of vehicles configured to use CNG. The transportation industry has indicated that LNG has greater applicability to long-haul trucking, rail locomotives, and shipping due to the size of the fuel tanks required, the energy density of the fuel, and the limited number of refueling stations available currently and expected in the near term. The growth in construction of these facilities has been concentrated along key interstate highways. Access to fueling locations and the cost of required onboard technology are seen as the most limiting factors to the expanded use of CNG in automobiles.

Some have called for increasing the pace of conversion of the U.S. transportation system, but the need to construct new fueling infrastructure is a key factor delaying widespread adoption. Some newer fueling locations are being built with capacity to allow private access as a means of encouraging greater use by non-fleet vehicles. However, early adoption in the United States has been largely in public fleets such as urban passenger buses, taxis, refuse haulers, and other vehicles that typically return each day to a base of operations that is, in most cases, strategically located near a fueling facility. Owners of large fleets make vehicle-purchasing decisions based on identified policy initiatives to encourage conversion or, if a policy is not clear, on favorable economics. An emerging trend is the public policy promotion of CNG in automobiles to incentivize the demand for fueling station construction through various initiatives involving tax incentives, public grants, and state fleet vehicle purchases. Public Clean Cities Coalition programs also provide support to fleet owners in urban areas considering whether to transition to natural gas for transportation.[27]

Gas can also be chemically reformulated, typically into energy-dense, liquid products. This approach can be classified either as gas-to-liquids (GTL) using the Fischer-Tropsch process or as the development of drop-in fuels that can directly substitute for conventional liquid fuels and distributed through traditional fueling stations. Fischer-Tropsch, although utilized for decades, is a large-scale, centralized, industrial process that requires capital expenditures in the billions of dollars. Sasol Limited, a South African company that was one of the commercial pioneers of the process, is planning a $14 billion plant in Louisiana that will be the second-largest GTL plant in the world, and will produce a synthetic diesel fuel compatible with fuel distribution infrastructure currently in place. This investment is a direct outcome of shale gas resources under development in the United States.[28]

There is ongoing research, both in academic institutions and industry, to reduce the scale of this process and make it more portable. Such an outcome has the potential to lead to the creation of synthetic fuels nearer to sources of shale gas, thus adding value to the commodity and producing a product that could be transported through means other than pipelines. In areas with smaller regions of stranded gas, a higher-value product with increased market demand could be developed. In a similar fashion, there is complimentary research exploring new chemical conversion processes independent of Fischer-Tropsch. Several companies, such as Primus Green Energy in New Jersey, have publicly announced the production of liquid fuels utilizing emerging technology, although on a pilot-scale level.[29] Researchers at several universities have produced dimethyl ether (DME) as a liquid from shale gas that has enhanced energy properties and a reduced environmental footprint.

CONCLUDING OBSERVATIONS

Natural gas development and production from shale formations is relatively new, but has transformed the world of energy. Many observers, including U.S. government officials, have only recently recognized the tremendous resource size and the benefits that may accrue from developing these resources, as well as the potential risks. Industry has improved its efficiency in extracting these resources, contributing to the sense that the world has moved from resource scarcity to resource abundance in hydrocarbons.

The change in U.S. natural gas supplies has raised the prospect of energy independence and possibly the United States becoming an exporter. This reversal of fortune of supply has sparked the interest of many other countries to emulate the U.S. success. However, no other country has been able to achieve the same level of transformation as has the United States. This is primarily because of the industry structure in the United States and mineral rights laws and regulations.

End Notes

[1] U.S. Energy Information Administration, Annual Energy Outlook 2014 with Projections to 2040, DOE/EIA0383(2014), Washington, DC, http://www.eia.gov/forecasts/aeo/pdf/0383(2014).pdf.

[2] Bentek Energy, LLC, Power Jump-Starts New Gas Market Cycle, December 2012.

[3] Bentek Energy, LLC, Power Jump-Starts New Gas Market Cycle, December 2012.

[4] WTI has been trading at a discount to Brent because of transportation constraints in multiple parts of the United States and other U.S. market variables. For additional analysis of on crude by rail in the United States, see CRS Report R43390, U.S. Rail Transportation of Crude Oil: Background and Issues for Congress, by John Frittelli et al.

[5] Coleman, James L., Robert C. Milici, Troy A. Cook, Ronald R. Charpentier, Mark Kirschbaum, Timothy R. Klett, Richard M. Pollastro, and Christopher J. Schenk, 2011, USGS, Assessment of Undiscovered Oil and Gas Resources of the Devonian Marcellus Shale of the Appalachian Basin Province.

[6] ICF International, Recently Published Assessments of India's Shale Gas Potential—Are They Realistic?, March 2012; home page of Terry Engelder, http://www3.geosc.psu.edu/~jte2/, last accessed on February 25, 2013.

[7] Pennsylvania Oil and Gas Reporting system data are available at https://www.paoiland gasreporting.state.pa.us/ publicreports/Modules/Production/ProductionHome.aspx.

[8] Credit Suisse, The Shale Revolution, December 13, 2012.

[9] Interstate Natural Gas Association of America, Natural Gas Pipeline and Storage Infrastructure Projections Through 2030, January 2009.

[10] Platts, "REX Gas Flows Drop 33% this Month Due to Marcellus, Spreads West," October 24, 2012.

[11] Bentek Energy, LLC, An Energy Market Fundamentals Report, November 2012.

[12] Bentek Energy, LLC, Power Jump-Starts New Gas Market Cycle, November 2012.

[13] National Fuel Gas Summary Annual Report, Your Future, Our Future, January 9, 2012.

[14] U.S. Department of Energy, Office of Fossil Energy, Long Term Applications Received by DOE/FE to Export Domestically Produced LNG from the Lower-48 States, June 11, 2014, http://energy.gov/sites/prod/files/2014/06/f16/Summary%20of%20LNG%20Export%20Applications.pdf.

[15] Forbes, "Dow Chemical Chief Wants to Limit U.S. LNG Exports," March 8, 2012, available at http://www.forbes.com/sites/christopherhelman/2012/03/08/dow-chemical-chief-wants-to-limit-u-s-lng-exports.

[16] Manufacturing & Technology News, "Gas Boom Is Fueling Manufacturing Resurgence," October 12, 2012, pp. 6-7.

[17] Sasol Inc, "Sasol Commences the Front-End Engineering and Design (FEED) Phase for an Integrated Gas-toLiquids," December 2, 2012; Platts, "Manufacturers, Producers See Different Futures for U.S. Natural Gas Supplies," August 22, 2012, available at http://www.platts.com/RSSFeedDetailedNews/RSSFeed/NaturalGas/6582012.

[18] Platts, "Manufacturers, Producers See Different Futures for U.S. Natural Gas Supplies," August 22, 2012, available at http://www.platts.com/RSSFeedDetailedNews/ RSSFeed/ NaturalGas/6582012.

[19] U.S. Energy Information Administration, "Gulf of Mexico Fact Sheet," available at http://www.eia.gov/special/ gulf_of_mexico/.

[20] International Energy Agency, Golden Rules for a Golden Age of Gas, World Energy Outlook, November 12, 2012.

[21] For additional information about European alternatives to Russian natural gas imports, see CRS Report R42405, Europe's Energy Security: Options and Challenges to Natural Gas Supply Diversification, coordinated by Michael Ratner.

[22] Financial Times, "Gazprom Reduces Price of Gas to Poland," November 6, 2012.

[23] Federal Register, "Dominion Cove Point LNG, LP; Application for Blanket Authorization to Export Previously Imported Liquefied Natural Gas," September 21, 2011, available at https://www.federalregister.gov/articles/2011/09/ 21/2011-24225/dominion-cove-point-lng-lp-application-for-blanket-authorization-to-export-previously-imported.

[24] Sierra Club, "Sierra Club Denies Permission to Build Liquefied Natural Gas Export Terminal," press release, April 26, 2012, http://sierraclub.typepad.com/compass/2012/04/ defeats-lng.html.

[25] Spectra Energy, "Spectra Energy Receives Notice to Proceed with Construction on New Jersey-New York Expansion Project," June 29, 2012.

[26] Times-Tribune, "Proctor and Gamble to Be 100% Energy Self-Sufficient by February," October 11, 2012, available at http://thetimes-tribune.com/news/business/procter-gamble-to-be-100-percent-energy-self-sufficient-by-february1.1386336.

[27] U.S. Department of Energy, Energy Efficiency and Renewable Energy, Building Partnerships to Reduce Petroleum Use in Transportation, available at http://www1.eere.energy.gov/ cleancities.

[28] Sasol Inc., "Sasol Commences the Front-End Engineering and Design (FEED) Phase for an Integrated Gas-toLiquids," December 2, 2012; Platts, "Sasol Advances Louisiana GTL-Cracker Complex," December 4, 2012, available at https://online.platts.com/PPS/P= m&e=1354665797920.5659393665237333516/GD_20121204.xml?artnum= cb76aa877-51c6-466d-8bf7-f0eb3a348b2c_3.

[29] IC Green Energy, Primus Green Energy, April 2011.

In: Producing Oil and Natural Gas …
Editor: Joelle Bolton ISBN: 978-1-63482-126-1

Chapter 3

POTENTIAL BUDGETARY EFFECTS OF IMMEDIATELY OPENING MOST FEDERAL LANDS TO OIL AND GAS LEASING*

Kathleen Gramp and Jeff LaFave

NOTE

Unless otherwise noted, all years are federal fiscal years, which run from October 1 to September 30. Numbers in the text and table may not add up to totals because of rounding.

The federal government offers private businesses the opportunity to bid on leases for the development of on- and offshore oil and natural gas resources on federal lands—although not all federally controlled lands are open to leasing now. The Congressional Budget Office (CBO) estimates that, under current laws and policies, the government's gross proceeds from all federal oil and gas leases on public lands will total about $150 billion over the next decade. CBO has analyzed a proposal to immediately open most federal lands to oil and gas leasing, which would affect the amounts the federal government collects in various fees and royalties both in the near term and over a longer period.

---

* This is an edited, reformatted and augmented version of a report issued by the Congressional Budget Office, August 2012.

Implementing such a proposal would open two categories of property now closed to development:

- Lands where leasing is now statutorily prohibited, notably, the Arctic National Wildlife Refuge (ANWR) and
- Onshore and offshore areas that are unavailable for leasing under current administrative policies, including sections of the Outer Continental Shelf (OCS)— generally, the submerged lands between 3 miles and 200 miles from the Atlantic, Pacific, and Florida coastlines—and certain onshore areas in which oil and gas leasing is either restricted or temporarily prohibited.

CBO expects that opening ANWR to development would yield about $5 billion in additional receipts over the next 10 years, primarily in the form of bonus payments made by private firms for the opportunity to explore for and develop resources in particular areas.

Most legislative proposals related to ANWR have specified that a significant portion of those receipts would be conveyed to the state of Alaska. Because extraction is currently prohibited, the receipts from leasing in ANWR could not be realized under current law, and any federal receipts that were projected to result from the change would be added to CBO's baseline estimates of collections from oil and gas development. The federal government also would collect royalties if oil and natural gas eventually were produced from those lands, but most royalty payments would not be collected until much later because of the long lag between the initial leasing agreement and the time when production begins.

According to estimates of potential resources by the Department of Energy's Energy Information Administration (EIA) and taking into account a range of probable oil prices, gross royalties from leasing in ANWR would probably total between $25 billion and $50 billion (in 2010 dollars) during the 2023–2035 period, or roughly $2 billion to $4 billion a year. (By comparison, CBO estimates that under current law, gross receipts from all federal oil and gas leasing activities in 2022 will be about $12 billion, in 2010 dollars.) The projected royalties from leasing in ANWR are very uncertain, however, as they depend both on the amount of oil that might be produced and on future oil prices. Any royalties collected from development in ANWR would be divided between Alaska and the federal government according to a formula that would be set by the authorizing legislation.

CBO anticipates that new legislation directing the Department of the Interior (DOI) to immediately offer most other federal lands for oil and gas leasing without any restrictions also would lead to an increase in federal receipts over the next decade. Specifically, with expanded leasing, CBO estimates that additional gross proceeds from federal oil and gas leases on public lands—principally in certain sections of the OCS off the Atlantic and Pacific coasts and in the eastern Gulf of Mexico and in onshore areas where leasing is now restricted—would total about $2 billion over the 2013–2022 period. (Most of that revenue is expected to come from the OCS leases; a portion of the proceeds would be shared with state governments.)

The long-term budgetary consequences of opening other federal lands to leasing are less clear, however. Much of the near-term development enabled by the proposal (beyond that in ANWR) would occur under current law, albeit at a later time. CBO does not have enough information to predict with specificity what would occur after 2022 either under current law or under the proposal. Under the proposal, income from royalties might be greater over the 2023–2035 period and smaller in subsequent years than under current law. But the proposal also might reduce the amount of bonus payments received between 2023 and 2035 because some of them would be collected sooner. Such long-term predictions are clouded by the inherent uncertainty surrounding market prices for oil and natural gas, state and local policies regarding resource development, and the potential impact of changes in technology.

Oil and Natural Gas Resources on Federal Lands

The budgetary effects of increasing the oil and gas industry's access to federal lands would depend on the quantity, characteristics, and market value of the untapped resources in the designated areas. CBO has relied on DOI's most recent geologic estimates of the quantity and quality of resources in forming its projections of revenue that would result from the proposed change in leasing.[1] However, estimates of oil and gas resources are by their nature inexact, particularly for areas that have not been explored or recently studied. As a result, DOI routinely revises its estimates of resources (sometimes upward and sometimes downward) as new data become available.

CBO's projections are based on DOI's mean estimate of undiscovered, technically recoverable resources—but CBO excluded most of DOI's estimates of natural gas resources in Alaska because gas extraction is not economically viable without a pipeline to transport the product to domestic or foreign markets. Using DOI's resource estimates and making that adjustment, CBO estimates that about 175 billion barrels of oil equivalent (BOE) exists in undiscovered oil and gas reserves on federal lands (excluding most of the natural gas reserves in Alaska)—nearly half of it in the central and western parts of the Gulf of Mexico (see Figure 1).[2] About 70 percent of the undiscovered oil and gas is under federal control on lands that are currently open to leasing; thus, additional receipts would come from opening the other 30 percent to leasing and production.

LEASING OFFSHORE FEDERAL LANDS

The geographic scope of leasing on the Outer Continental Shelf has changed often over the past few decades.[3] CBO anticipates that, under current law, DOI will offer leases for most of the acreage in the OCS over the next several decades.

Until the early 1980s, DOI offered leases in all of the OCS, including the areas off the Atlantic, Pacific, and Florida coasts. In 1990, after the Congress imposed a series of temporary restrictions, President George H.W. Bush withdrew large portions of the OCS in the Atlantic and Pacific Oceans and the eastern Gulf of Mexico from the leasing program. Those restricted areas were subsequently expanded by President Clinton. Then, in 2008, President George W. Bush narrowed the restrictions to include only areas that had been designated as National Marine Sanctuaries. In 2010, President Obama removed Alaska's Bristol Bay area from the leasing program until the end of June 2017.

Since 2008, policies on leasing in the Atlantic and Pacific OCS have varied, reflecting differences between the two most recent Administrations. In January 2009, DOI issued a proposed five-year plan that included lease sales in the Atlantic and Pacific OCS for the 2010–2015 period. The program proposed in June 2012 does not include an option for sales in those areas between 2012 and 2017. Neither plan involved the areas in the Gulf of Mexico adjacent to the Florida coast in which leasing is now prohibited until the end of June 2022.[4]

Other than the temporary ban on leasing in the eastern Gulf of Mexico, there currently are no *statutory* restrictions on OCS leasing. Decisions about leasing are made administratively—in consultation with industry and the states—for five-year periods. Leases cannot be offered for areas that are not included in a five-year plan, but the available regions may change whenever a new plan is adopted. The next plan is expected to go into effect in August 2012 and will extend for five years unless a future Administration chooses to restart the process before that plan expires.

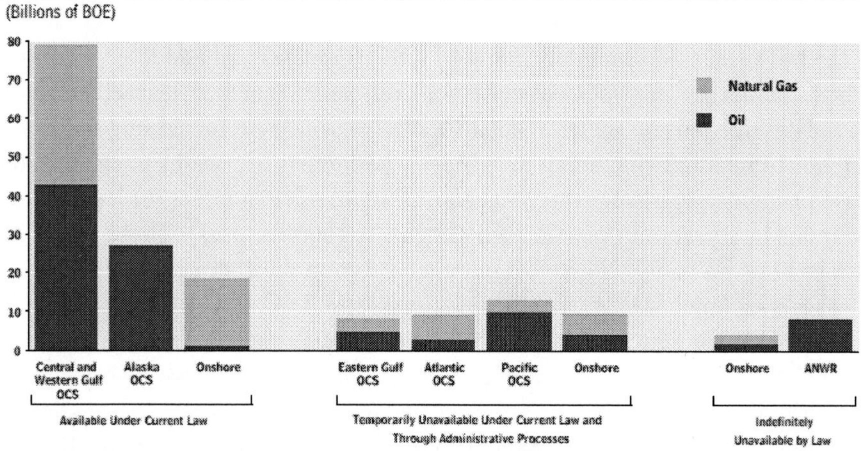

Source: Congressional Budget Office using data from the Department of Agriculture, U.S. Forest Service; the Department of Energy, Energy Information Administration; and the Department of the Interior, U.S. Geological Survey, Bureau of Land Management.

Notes: BOE (barrel of oil equivalent) is a measure of the approximate amount of energy released by burning one barrel (42 U.S. gallons) of crude oil; 1,000 cubic feet of natural gas = 0.178 BOE.

The areas of the Outer Continental Shelf (OCS) referenced here are the submerged lands generally between 3 miles and 200 miles from the Atlantic and Pacific coasts of the continental United States, the coast of Alaska, and the coast of the Gulf of Mexico. ANWR is the Arctic National Wildlife Refuge in Alaska.

Figure 1. Undiscovered Oil and Gas Resources on Federal Lands, by Location.

Historical experience suggests that only a fraction of the leases awarded in the OCS will eventually be brought into production. Almost 60 percent of the OCS leases issued in the Gulf of Mexico through 2007 either expired or were relinquished without producing any oil or natural gas.[5] CBO estimates that almost 90 percent of the 2011 OCS production was from leases issued before 2001, reflecting the long lead times associated with exploring and developing oil and gas fields.[6]

Leasing Onshore Federal Lands

On the basis of information in the 2008 interdepartmental inventory of onshore resources and in DOI's 2010 updated estimates of undiscovered oil and gas resources in Alaska, CBO estimates that roughly 60 billion BOE of oil and natural gas resources are located under federal lands (excluding ANWR, which is discussed separately below).[7] CBO estimates that about 80 percent of those resources are located under federal lands that are leased, currently available for leasing under standard terms, or available for leasing subject to minor stipulations (including temporary withdrawals of various tracts for land-use planning, seasonal restrictions on drilling that are in effect for less than six months in a year, and certain requirements for surface uses). Those terms and constraints probably will have a minimal effect on the commercial value of the leases over time.[8]

About 15 percent of onshore resources are covered by administrative prohibitions on leasing or are subject to major stipulations (including prohibitions on drilling on the surface directly above the leased tract and seasonal restrictions on drilling for more than six months in a year) that could significantly affect the lease values. The other 5 percent of the onshore area, which includes national parks and wilderness areas, is by statute closed to leasing.

Information from DOI indicates that most administrative restrictions and lease stipulations governing federal onshore oil and gas resources are aimed primarily at protecting plants and wildlife. Federal agencies can impose or lift those restrictions at any time, so their long-term effects on leasing, production, and federal receipts are not quantifiable. However, CBO expects that, under current law, over the next several decades most onshore oil and gas resources will be offered under lease terms that would not significantly influence the likelihood of companies' bidding for leases or change the leases' economic value.

CBO's Baseline Projections of Receipts from Oil and Gas Leasing

CBO estimates that, under current laws and policies, gross proceeds from all federal oil and natural gas leases on public lands will total about $150 billion over the 2012–2022 period (see Table 1); some of those receipts will be shared with the states. Royalties, which are assessed on the value of oil and gas produced, are projected to account for about 80 percent of the total. (That projection is based on CBO's forecast of oil and gas prices and on information from DOI, EIA, and various industry sources that project future production.)

Under current policies, lessees pay a royalty rate of 18.75 percent for new OCS leases or 12.5 percent for new onshore leases.[9] In 2011, roughly 20 percent of OCS production was exempt from royalties under the OCS Deep Water Royalty Relief Act.[10] Lessees also make annual rental payments before a lease goes into production. CBO's projections of rental income are based on estimates of the volume of nonproducing acreage multiplied by the per-acre fee specified for each lease sale.

Table 1. CBO's March 2012 Baseline Projections of Gross Proceeds from Federal Oil and Gas Leasing, by Fiscal Year

(Billions of dollars)

	2012	2013	2014	2015	2016	2017	2018	2019	2020	2021	2022	Total, 2012–2022
					Offshore Leases[a]							
Bonus Payments	1.7	1.8	1.3	1.4	1.6	1.4	1.4	5.6	2.3	2.0	1.3	21.8
Royalties and Rents	5.7	6.1	6.7	7.2	7.9	7.9	9.3	9.5	10.1	11.0	11.7	93.3
Subtotal	7.4	7.9	8.0	8.6	9.5	9.3	10.7	15.1	12.4	13.0	13.0	115.1
					Onshore Leases							
Bonus Payments	0.2	0.2	0.2	0.2	0.2	0.2	0.2	0.2	0.2	0.2	0.2	2.2
Royalties and Rents	2.4	2.5	2.6	2.7	2.8	2.9	3.0	3.1	3.1	3.2	3.3	31.6
Subtotal	2.6	2.7	2.8	2.9	3.0	3.1	3.2	3.3	3.3	3.4	3.5	33.8
Total	10.0	10.6	10.8	11.5	12.5	12.4	13.9	18.4	15.7	16.4	16.5	148.9
Memorandum:												
CBO's March 2012 Economic Assumptions												
Oil price[b]	101.1	98.4	94.2	92.8	95.2	97.9	100.8	103.8	106.9	110.1	113.5	n.a.
Natural gas price[c]	3.6	4.2	4.6	5.0	5.2	5.3	5.4	5.5	5.7	5.8	5.9	n.a.

Source: Congressional Budget Office.

Notes: Gross receipts are the amounts paid by lessees, including the portion the federal government will share with states. n.a. = not applicable.

[a] Offshore leases consist of those in areas of the Outer Continental Shelf generally between 3 miles and 200 miles from the Atlantic and Pacific coasts of the continental United States and in the eastern Gulf of Mexico.

[b] Average refiner acquisition cost in dollars per barrel (42 U.S. gallons).

[c] Price at the Henry hub in dollars per thousand cubic feet.

Firms also pay a bonus when they acquire a lease that allows the winning bidder an opportunity to develop oil and gas resources within a specified period. CBO's estimates of the bonus payments are based on historical trends in lease sales in areas that are likely to be available for leasing over the next 10 years. Proceeds from bonus payments vary from year to year, depending on bidders' assessments of the potential profitability of the acreage being offered in each auction. Bonus bids for OCS leases spiked in fiscal year 2008, for example, when DOT held a special lease sale for areas in the Chukchi Sea in Alaska and three sales for areas in the Gulf of Mexico, which included previously awarded deep-water leases that had expired when they were not brought into production within a 10-year term. CBO anticipates that bonus payments will increase again in 2019 when DOT auctions expiring leases from the 2008 sales. CBO's baseline projections also reflect the possibility that DOT will offer leases in the Atlantic and Pacific OCS after 2017.

CBO updates its baseline projections annually to incorporate new information on potential production and commodity prices, including changes that result from new technology. Advances in the methods for extracting natural gas from shale formations have dramatically increased the domestic supply of gas, in turn lowering wellhead prices and thus reducing federal royalties for natural gas produced in the OCS. Most of the nation's known shale gas resources are on nonfederal lands, and it is unclear whether the reduction in royalties attributable to lower prices will be offset over the next 10 years by an increase in gas production on federal lands.

Other advances are expected to boost receipts: Most of the growth in production from the Gulf of Mexico is projected to come from discoveries in deep water that were not technically or economically viable in the past. Similarly, CBO's baseline projections take account of the possibility that it will be feasible to develop oil shale reserves by 2022, thus boosting receipts from onshore leasing.

It is important to note, however, that any projection that involves geologic resources is inherently uncertain. In the 1970s and 1980s, bidders paid more than $2.8 billion ($7.7 billion in 2010 dollars) for leases in the Atlantic OCS that turned out to be geologically unproductive or too expensive to produce. In other instances, firms have discovered large quantities of oil and gas on leases they acquired relatively inexpensively.[11] The wide variation in such results underscores the riskiness of oil and gas investments and the difficulty of predicting whether or when production will be possible in any given area.

PROJECTIONS OF OIL PRODUCTION AND FEDERAL RECEIPTS FROM ANWR LEASES

If the statutory ban on leasing in ANWR—which is estimated to contain roughly 8 percent of the nation's undiscovered oil (see Figure 1 on page 3)—was lifted, significant new opportunities for oil production would become available. ANWR is believed to contain a particularly large volume of oil (about 8 billion barrels) in a relatively small area (roughly 1.5 million acres), and those resources could be simpler and less expensive to develop than is the case for some other areas.[12] In contrast to ANWR, other onshore federal lands are believed to hold about 8 billion barrels of undiscovered oil, dispersed over some 280 million acres.[13]

CBO estimates that bonus payments from leasing in ANWR would increase gross federal receipts by $5 billion over the 2013–2022 period.[14] Under current law, 90 percent of that money would be paid to the state of Alaska and 10 percent would be deposited in the U.S. Treasury. Most legislative proposals related to ANWR that have been introduced over the past two decades have called for 50 percent of bonus payments and royalties to go to the federal government and 50 percent to the state.

If legislation was enacted in 2013 to open ANWR to leasing, no production would be likely to occur for 10 years and production probably would not peak before 2032. The federal government would receive no royalties from those leases until production began.

Forecasts of energy prices over 20 years are not very reliable, and they usually encompass a wide range. Assuming that oil prices over the 2023–2035 period might range from under $100 per barrel to over $150 per barrel (in 2010 dollars), and using data published by EIA about the amount of oil that could be produced from ANWR, CBO estimates that the government's gross receipts from royalties might total between $25 billion and $50 billion (in 2010 dollars) over that period.[15] The federal portion of the royalties could be as high as $25 billion (if 50 percent went to Alaska) or as low as $2.5 billion (if 90 percent went to Alaska).

PROJECTIONS OF OIL AND GAS PRODUCTION AND FEDERAL RECEIPTS FROM LEASING OUTSIDE ANWR

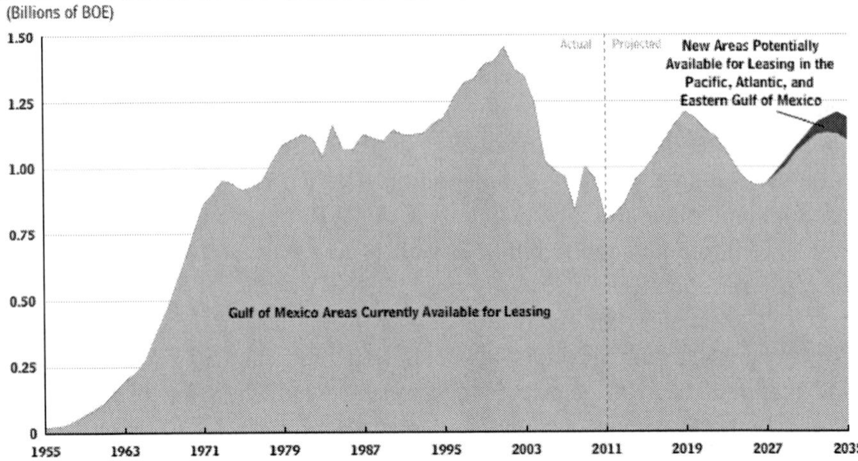

Source: Congressional Budget Office using data from the Department of Energy, Energy Information Administration (EIA); and the Department of the Interior, Bureau of Ocean Energy Management.

Notes: Historical data on production, which began in 1947, are from the Bureau of Ocean Energy Management, and projections for the 2012–2035 period are based on data from EIA.

BOE (barrel of oil equivalent) is a measure of the approximate amount of energy released by burning one barrel (42 U.S. gallons) of crude oil; 1,000 cubic feet of natural gas = 0.178 BOE.

Figure 2. History and EIA Projections of Oil and Gas Production from Existing and Proposed Offshore Leasing Areas.

Where and when the government offers leases for areas other than ANWR over the next 10 years will depend on administrative and legislative actions. If policymakers initiated extensive leasing in new areas, additional gross proceeds from federal oil and gas leases on public lands— including the OCS regions of the Atlantic and Pacific Oceans and the eastern Gulf of Mexico and onshore areas where leasing is now restricted—would total about $2 billion over the 2013–2022 period, CBO estimates.[16] Most of that revenue would come from OCS leases. CBO expects that most of the added production would occur after 2022 because of the time needed to prepare for auctions, explore and develop oil and gas fields, and obtain the necessary state and local permits for processing and marketing oil and gas.

Offshore Leasing

For this analysis, CBO used EIA's estimates of the potential for new areas to produce oil or gas after 2022. EIA expects that any initial production from newly opened areas in the Atlantic, Pacific, and eastern Gulf of Mexico would be far less than is produced by current operations in the Gulf of Mexico (see Figure 2). In its *Annual Energy Outlook 2011*, EIA estimated that if leasing commenced in those OCS regions by 2023, production through 2035 would amount to around 0.35 billion BOE—or about 3 percent of the 13.5 billion BOE that the agency projected would be produced from federal leases in the Gulf of Mexico over that 13-year period.[17]

EIA's estimates reflect its assumption that "local infrastructure issues and other potential nonfederal impediments are resolved."[18] In CBO's view, such factors probably would slow or limit production, as they sometimes have in the past. The federal government has spent about $1.5 billion to compensate firms for leases that were canceled or relinquished because of state or local concerns about oil and gas development off the coasts of California, North Carolina, and Florida and in Bristol Bay in Alaska.[19] According to DOI, 24 localities in California have "enacted ordinances that either bar the construction of onshore support facilities for offshore oil and gas development or subject the approval of such facilities to a vote by local citizens."[20] Any development in the Atlantic OCS would involve siting and building new pipelines and related onshore facilities, which would require approval by state and local authorities.

Other technical complications and economic factors add to the uncertainty surrounding forecasts of production in new areas of the OCS. DOI's resource assessments suggest that much of the undiscovered oil in the eastern Gulf of Mexico is located in ultradeep water—water that is more than 2,400 meters (about 7,900 feet) deep—where few leases can be brought into production in any year because of the cost and complexity of their development.[21] Other factors could slow production in new areas, including the need for exploratory drilling and the expectation that most of the fields will be relatively small.[22] Historically, production facilities have been installed at a slower pace in the Reflecting EIA's projections and the considerations discussed above, CBO anticipates that production from new offshore areas will be determined primarily by how policymakers in California would respond to the possibility of new oil and gas development. Production from the California OCS accounts for nearly 80 percent of the estimated 0.35 billion BOE projected by EIA for production over the 2023–2035 period; most of the remainder would come from new production of oil in the eastern Gulf of Mexico. EIA currently estimates that no oil or gas will be produced in the Atlantic OCS through 2035, in keeping with its assumptions that future oil and gas prices will be similar to those in the agency's forecast and that the region's geologic characteristics are as they were identified by DOI in 2006.

The uncertainty surrounding whether and when new offshore areas will be developed in the future makes it difficult to estimate the budgetary impact of accelerating leasing. If leasing started sooner than currently assumed by EIA—for example, by 2017 instead of 2023 for the California and Florida OCS—the net increase in royalties could range from an average of tens of millions dollars a year to a few hundred million dollars a year over the 2023–2035 period, depending on whether policymakers in California allowed the development of new offshore leases—which is very uncertain.[24] Such gains would diminish over time and eventually turn negative because the resources in those areas would be depleted sooner. Similarly, collecting bonus payments earlier could reduce the amount received during the 2023–2035 period, which could offset some of the estimated increase in royalties over that period.

Onshore Leasing

CBO cannot project receipts beyond 2022 for leases on federal lands other than ANWR because it lacks data to inform predictions about onshore resources—and by extension, production—on federal lands after that year. The information provided in this report regarding *offshore* oil and gas production after 2022 is based on EIA's projections. But that agency's projections of *onshore* oil and gas production do not include specific projections for federal lands.

Long-Term Budgetary Effects of Expanded Leasing

Whether future Administrations will make areas in the Atlantic, Pacific, or eastern Gulf of Mexico available for oil and gas leasing by 2023 is not known, but CBO expects that such leasing will occur over time without any changes in law. Legislation to require immediate leasing of those areas would accelerate development but probably would not affect the total amount of development in those areas over the next several decades.

End Notes

[1].Estimates of offshore resources are based on information in Department of the Interior, Bureau of Ocean Energy Management, Assessment of Undiscovered Technically Recoverable Oil and Gas Resources of the Nation's Outer Continental Shelf, 2011, BOEM Fact Sheet RED-2011-01a (November 2011), http://go.usa.gov/ wbD. Estimates of onshore resources are based on a joint report of the Departments of the Interior, Agriculture, and Energy, Inventory of Onshore Federal Oil and Natural Gas Resources and Restrictions to Their Development: Phase III Inventory—Onshore United States (2008), http://go.usa.gov/wQ0. CBO adjusted the 2008 resource assessments to include information in Department of the Interior, U.S. Geological Survey, 2010 Updated Assessment of Undiscovered Oil and Gas Resources of the National Petroleum Reserve in Alaska (NPRA), Fact Sheet 2010-3102 (October 2010), http://go.usa.gov/GDg.

[2].The barrel of oil equivalent is a measure of the approximate amount of energy released by burning one barrel (42 U.S. gallons) of crude oil; 1,000 cubic feet of natural gas equals 0.178 BOE.

[3]. See Adam Vann, Offshore Oil and Gas Development: Legal Framework, CRS Report for Congress RL33404 (Congressional Research Service, May 2, 2011); and Curry L. Hagerty, Outer Continental Shelf Moratoria on Oil and Gas Development, CRS Report for Congress R41132 (Congressional Research Service, March 23, 2011).

[4]. That prohibition was enacted in title I, division C of the Gulf of Mexico Energy Security Act of 2006; Public Law 109-432; 120 Stat. 3000, 3003.

[5] Department of the Interior, Bureau of Ocean Energy Management, Estimated Oil and Gas Reserves, Gulf of Mexico OCS Region, December 31, 2007, OCS Report BOEMRE 2011-045 (September 2011), p. 13, http://go.usa.gov/wPX.

[6] For production estimates, see Bureau of Safety and Environmental Enforcement, Federal OCS Oil and Gas Production (June 2012), http://go.usa.gov/wUg; and Bureau of Ocean Energy Management, "Production Data Online Query" (accessed August 9, 2012), http://go.usa.gov/GvM.

[7] That resource estimate includes undiscovered, technically recoverable resources; estimated growth in reserves; and proved reserves.

[8] Timothy Fitzgerald, "Evaluating Split Estates in Oil and Gas Leasing," Land Economics, vol. 86, no. 2 (May 2010), pp. 294–312. Royalties usually are set as a percentage of the value of production. At those rates and a wellhead price of $100 per barrel for oil, the government's gross royalty would be $18.75 or $12.50 per barrel of oil produced from a new offshore or onshore lease, respectively. If the wellhead price of natural gas was $3 per thousand cubic feet, the gross federal royalty would be 56 cents or 38 cents per thousand cubic feet for natural gas produced from a new offshore or onshore lease, respectively, or about $3 or $2, respectively, per BOE in natural gas. How much the federal government receives from royalties depends on how much is shared with state governments and on how much of the oil and gas produced is eligible for royalty relief.

[10] 43 U.S.C. § 1337 (2006 & Supp.).

[11] For example, according to DOI, $11 million (or about $18 million in 2010 dollars) was paid for leases for the Mars–Ursa fields in the Gulf of Mexico, which are now estimated to contain more than 1.5 billion BOE.

[12] Department of the Interior, U.S. Geological Survey, Arctic National Wildlife Refuge, 1002 Area, Petroleum Assessment, 1998, Including Economic Analysis USGS Fact Sheet FS-028-01 (April 2001), http://go.usa.gov/wPN.

[13] Departments of the Interior, Agriculture, and Energy, Inventory of Onshore Federal Oil and Natural Gas Resources and Restrictions to Their Development: Phase III Inventory—Onshore United States (2008), http://go.usa.gov/wQ0; the largest field in ANWR is estimated to be about one-tenth the size of Alaska's Prudhoe Bay field, the largest oil field in North America.

[14] See Congressional Budget Office, cost estimate for H.R. 3407, the Alaskan Energy for American Jobs Act, (February 7, 2012).

[15] Department of Energy, Energy Information Administration, Annual Energy Outlook 2012: With Projections to 2035, DOE/ EIA-0383(2012) (June 2012), http://go.usa.gov/GjA, and Analysis of Crude Oil Production in the Arctic National Wildlife Refuge, SR/ OIAF/2008-03 (May 2008), http://go.usa.gov/GW9.

[16] See Congressional Budget Office, cost estimate for H.R. 3410, the Energy Security and Transportation Jobs Act (February 7, 2012).

[17] This estimate is based on the analysis in Department of Energy, Energy Information Administration, Annual Energy Outlook 2011: With Projections to 2035, DOE/EIA-0383(2011) (April 2011), http://go.usa.gov/wPV. That report discusses the impact on production from "lower-48 offshore" areas that resulted from reinstating limits on leasing in the Atlantic and Pacific OCS and in the eastern Gulf of Mexico. Those effects are measured relative to the amounts projected in EIA's "reference" case. EIA's Annual Energy Outlook 2012 includes updated estimates for OCS production but does not provide estimates of the impact of reimposing limits on OCS leasing.

[18] Statement of Howard Gruenspecht, Acting Administrator, Energy Information Administration, before the Subcommittee on Energy and Mineral Resources, House Committee on Natural Resources (March 5, 2009), p. 8, http://go.usa.gov/w0N.

[19] See Amber Res. Co. v. United States, 538 F.3d 1358 (Fed. Cir. 2008); Mobil Oil Exploration v. United States, 530 U.S. 604 (2000); Department of the Interior, U.S. Fish and Wildlife Service, "Interior Reaches Agreement to Acquire Mineral Rights in Everglades, Settles Litigation on Offshore Oil and Gas Leases in Destin Dome" (press release, May 29, 2002), http://go.usa.gov/ w0R; and Department of the Interior, "Landmark Protections Announced for Fragile Offshore Resources" (press release, July 31, 1995), http://go.usa.gov/w0n.

[20] Department of the Interior, Minerals Management Service, Report to Congress: Comprehensive Inventory of U.S. OCS Oil and Natural Gas Resources (February 2006), p. 100, http://go.usa.gov/wb9.

[21] See Department of the Interior, Bureau of Ocean Energy Management, "2006 Assessment Results Data and Spreadsheets: UTTR by Water Depth by Planning Area," Assessment of Undiscovered Technically Recoverable Oil and Gas Resources of the Nation's Outer Continental Shelf, 2006 (February 2006), http://go.usa.gov/G4c. According to DOI's leasing data, only 2 percent of the active leases in areas where the water is deeper than 5,000 feet are in production, compared with about 10 percent in areas where the water is 1,000–5,000 feet deep and 50 percent in areas where the water is considered shallow.

[22] According to DOI, firms drilled more than 47,000 wells in the central and western Gulf of Mexico through 2007. By contrast, 64 exploratory wells have been drilled in the eastern Gulf and about 50 wells are in the Atlantic OCS. See Department of the Interior, Bureau of Ocean Energy Management, Estimated Oil and Gas Reserves, Gulf of Mexico Region, December 31, 2007, OCS Report BOEMRE 2011-045 (September 2011), p. 13, http://go.usa.gov/wPX; and Department of the Interior, Minerals Management Service, Report to Congress: Comprehensive Inventory of U.S. OCS Oil and Natural Gas Resources (February 2006), pp. 84–85, http://go.usa.gov/wb9. For information on field sizes, see

Department of Energy, Energy Information Administration, Analysis and Projections, "Oil and Gas Supply Module (OGSM)," Annual Energy Outlook—Model Documentation, p. 3-3 (July 2011), http://go.usa.gov/wPA.

[23] DOI issued 470 leases for the Pacific OCS from1963 through 1984; 43 were brought into production. Over that period, 23 production facilities were installed, but no more than 3 were added in any year. By contrast, more than 100 facilities were installed each year in the Gulf of Mexico during the 1960s and 1970s. See Bureau of Ocean Energy Management, "Installations and Removals—Offshore Production Facilities in Federal Waters" (accessed August 9, 2012), http://go.usa.gov/wPo.

[24] EIA's estimates reflected the assumption that leasing in the South and Mid-Atlantic regions would begin in 2018. The state of California currently opposes new offshore oil and gas development. See West Coast Governors Alliance on Ocean Health, "Comments on BOEMRE's 5-Year Outer Continental Shelf Oil and Gas Leasing Program: 2010–2015," Governors' Letters and Replies: Offshore Oil and Gas (August 29, 2008), www.westcoastoceans.org/index.cfm?content.display&pageID= 136#OilGas.

INDEX

D

E

F

G

M

N

O

P